Springer Complexity

Springer Complexity is an interdisciplinary program publishing the best research and academic-level teaching on both fundamental and applied aspects of complex systems – cutting across all traditional disciplines of the natural and life sciences, engineering, economics, medicine, neuroscience, social and computer science.

Complex Systems are systems that comprise many interacting parts with the ability to generate a new quality of macroscopic collective behavior the manifestations of which are the spontaneous formation of distinctive temporal, spatial or functional structures. Models of such systems can be successfully mapped onto quite diverse "real-life" situations like the climate, the coherent emission of light from lasers, chemical reaction-diffusion systems, biological cellular networks, the dynamics of stock markets and of the internet, earthquake statistics and prediction, freeway traffic, the human brain, or the formation of opinions in social systems, to name just some of the popular applications.

Although their scope and methodologies overlap somewhat, one can distinguish the following main concepts and tools: self-organization, nonlinear dynamics, synergetics, turbulence, dynamical systems, catastrophes, instabilities, stochastic processes, chaos, graphs and networks, cellular automata, adaptive systems, genetic algorithms and computational intelligence.

The three major book publication platforms of the Springer Complexity program are the monograph series "Understanding Complex Systems" focusing on the various applications of complexity, the "Springer Series in Synergetics", which is devoted to the quantitative theoretical and methodological foundations, and the "SpringerBriefs in Complexity" which are concise and topical working reports, case-studies, surveys, essays and lecture notes of relevance to the field. In addition to the books in these two core series, the program also incorporates individual titles ranging from textbooks to major reference works.

Understanding Complex Systems

Founding Editor: S. Kelso

Future scientific and technological developments in many fields will necessarily depend upon coming to grips with complex systems. Such systems are complex in both their composition – typically many different kinds of components interacting simultaneously and nonlinearly with each other and their environments on multiple levels – and in the rich diversity of behavior of which they are capable.

The Springer Series in Understanding Complex Systems series (UCS) promotes new strategies and paradigms for understanding and realizing applications of complex systems research in a wide variety of fields and endeavors. UCS is explicitly transdisciplinary. It has three main goals: First, to elaborate the concepts, methods and tools of complex systems at all levels of description and in all scientific fields, especially newly emerging areas within the life, social, behavioral, economic, neuro- and cognitive sciences (and derivatives thereof); second, to encourage novel applications of these ideas in various fields of engineering and computation such as robotics, nano-technology and informatics; third, to provide a single forum within which commonalities and differences in the workings of complex systems may be discerned, hence leading to deeper insight and understanding.

UCS will publish monographs, lecture notes and selected edited contributions aimed at communicating new findings to a large multidisciplinary audience.

More information about this series at http://www.springer.com/series/5394

Ralph Kenna • Máirín MacCarron •
Pádraig MacCarron
Editors

Maths Meets Myths: Quantitative Approaches to Ancient Narratives

 Springer

Editors
Ralph Kenna
Applied Mathematics Research Centre
Coventry University
Coventry, United Kingdom

Máirín MacCarron
Department of History
University of Sheffield
Sheffield, United Kingdom

Pádraig MacCarron
Department of Experimental Psychology
University of Oxford
Oxford, United Kingdom

ISSN 1860-0832   ISSN 1860-0840   (electronic)
Understanding Complex Systems
ISBN 978-3-319-81887-0   ISBN 978-3-319-39445-9   (eBook)
DOI 10.1007/978-3-319-39445-9

Printed on acid-free paper

This Springer imprint is published by Springer Nature
The registered company is Springer International Publishing AG Switzerland

Foreword

It has long been accepted that the mathematical sciences are intrinsic to the understanding of our physical reality. This wonderfully diverse and consistently stimulating collection of essays takes the reach of maths a step further: it demonstrates that maths permeates the structures of our imaginative universe too. A distinctive attribute of this exciting volume is its place at the cutting edge of interdisciplinarity. It is perhaps unsurprising that its engagements between, for instance, network science and literary criticism, algorithms and medieval historiography and statistics and folklore studies—to mention only a few of the innovative pairings at work here—are only now breaking new ground. Scientists have always rightly been puzzled by how some who work in the humanities feel free to boast about their lack of mathematical skills and knowledge, while scientists who confess ignorance of literature, philosophy or history are shamed for it. As the Introduction to this volume explains, the "two cultures" have tended hitherto to keep carefully within their own spheres, and no wonder. But all of the contributors to *Maths Meets Myths* boldly cross old disciplinary boundaries, with a range far exceeding the eponymous confluence of mathematicians and mythographers. The results are an eye-opening glimpse into the future of academic research.

Linacre College   Heather O'Donoghue
University of Oxford   Professor of Old Norse and Vigfússon Rausing
Oxford, UK   Reader in Old Icelandic Literature and
                                                            Antiquities

Preface

In recent years, scientists from the Statistical Physics Group of the Applied Mathematics Research Centre[1], supported by national and international funding agencies, have been developing a new, quantitative approach to the analysis of ancient texts. Our first foray into this traditional humanities domain focused on comparative mythology, and our first journal publication on the topic generated worldwide interest and impact, including radio interviews and newspaper articles. That publication was in Europhysics Letters, the flagship journal of the European Physical Society. It is, of course, unusual to publish a paper about mythology in a journal devoted to physics, but statistical physicists have long been interested in outward applications, and the development of complex-network science over the past 20 years, in particular, has facilitated an explosion of interdisciplinary activity. Within a year, the paper became the most downloaded in the history of the journal, demonstrating an extraordinary resonance with a broad readership.

As our research developed, we became aware of excellent and highly original quantitative studies of myths, folktales and age-old chronicles being carried out by various individuals and small groups around the world. These include computational investigations of the narrative contents of ancient annals and revolutionary studies of folktales using methods as diverse as phylogenetics and principal component analysis, as well as other fascinating approaches. As our journey from our "native discipline" of theoretical physics into the humanities progressed, we also became aware that significant amounts of data have been gathered by humanities scholars and we think that these are amenable to new mathematical, statistical and computational approaches.

Supported by the European Science Foundation and other sources, we hosted workshops designed to bring like-minded people together from various academic disciplines. We wished to facilitate expert researchers and internationally renowned

---

[1]The Applied Mathematics Research Centre is a member of the so-called L^4 Collaboration involving statistical physics groups in Leipzig University, Germany; Lorraine University, France; the Institute of Condensed Matter Physics in Lviv, Ukraine; and Coventry University, England.

scholars from different disciplines learning from and collaborating with each other. Besides scientists with the right tools and interests, we invited people with knowledge of the humanities who also understand, or at least appreciate, the new quantitative approaches. Our aims thus expanded to the exploration of quantitative modelling for the understanding of textual narratives concerning past cultures in a broad sense.

We chose the alliterative title *Maths Meets Myths* to reflect the ostensible polarity of the contents of the workshops, although they extended beyond these two disciplines. The international *Maths Meets Myths* team now includes physicists, applied mathematicians, complexity theorists, computer scientists, anthropologists, psychologists, authors, filmmakers, artists, historians, medievalists and other scientists and humanities scholars. The workshops had about a 50:50 science–humanities balance. The remit was for the scientists to display their wares (quantitative tools) and the humanities people to present issues which may be susceptible to quantitative approaches. Participants were asked to try to keep their contributions as non-technical as possible as about half the audience was from the "other side". The idea was that this may aid the promotion of cooperation and collaboration.

This volume grew partly out of our workshops, but it is not a traditional set of proceedings. Instead, we selected contributions which, from a scientific point of view, were most firmly established or, from a humanities point of view, were most "ripe" for data harvesting. We also approached eminent academics who were not involved in the *Maths Meets Myths* workshops. Authors were asked to present contributions which included one or more of the following categories: (a) results of an application of state-of-the-art quantitative approaches to sources from the past (folktales, fables, myths, legends, sagas, epics and histories), (b) a description of a new quantitative method which has been applied elsewhere and could be applied to the sources we're interested in here and (c) a description of sources which would be amenable to future quantitative treatment and why they are important. Following the successful experience of our meetings, contributors were asked to be as non-technical as possible and to deliver chapters which are comprehensible to non-experts. All contributions were peer reviewed.

To scientists involved in this project, myths, folktales, chronicles and histories contain treasure troves of fascinating and often complex systems—vast amounts of data just waiting to be analysed quantitatively. To humanities experts, the scientists offer new and exciting ways to interrogate familiar sources, to explore old questions in a new light and to open new avenues of research. Our hope is that the outcome—in the form of this book—will well represent what science can offer to the humanities and vice versa. We also hope that this volume will help demonstrate the value of collaboration between the natural sciences and humanities and to help forge a community embracing the two.

Coventry, UK   Ralph Kenna
Sheffield, UK   Máirín MacCarron
Oxford, UK   Pádraig MacCarron

Contents

Introduction

Ralph Kenna, Máirín MacCarron, and Pádraig MacCarron

The relationship between science and the humanities has been the subject of a great deal of discussion and comment. Perhaps most notorious and controversial is the *two cultures* viewpoint advanced by Charles Snow in 1959. Snow lamented what he perceived as a gulf of ignorance between the two areas and argued that difficulties in communication between them pose a significant impediment to addressing many of the world's problems. Snow's diagnosis provoked an enormous amount of controversy and is still debated to this day.

The Two Cultures dichotomy is unfortunate, in our view, as science and the humanities are perhaps not so far apart as it pretends. Indeed, the spirit of curiosity that drives many scientists is the same one that motivates many activities in the humanities. Moreover, the two branches are part of the same ultimate endeavour— systematic investigations aimed at achieving greater understandings of the world in which we live. Their disciplines do not exist in isolation and they are nowadays increasingly interconnected and interdependent. For example, our knowledge of the past draws from modern scientific analyses of archaeological discoveries alongside

---

R. Kenna (✉)
Applied Mathematics Research Centre, Coventry University, Coventry CV1 5FB, UK

Doctoral College for the Statistical Physics of Complex Systems,
Leipzig-Lorraine-Lviv-Coventry (L^4), 04009 Leipzig, Germany
e-mail: r.kenna@coventry.ac.uk

M. MacCarron
Department of History, University of Sheffield, Sheffield S3 7RA, UK
e-mail: m.maccarron@sheffield.ac.uk

P. MacCarron
Department of Experimental Psychology, University of Oxford, South Parks Road, Oxford OX1 3UD, UK
e-mail: padraig.maccarron@wolfson.ox.ac.uk

© Springer International Publishing Switzerland 2017
R. Kenna et al. (eds.), *Maths Meets Myths: Quantitative Approaches to Ancient Narratives*, Understanding Complex Systems,
DOI 10.1007/978-3-319-39445-9_1

a range of historical approaches; anthropology imports notions and methods from the social sciences, as well as from evolutionary biology; sociology and economics require statistics and mathematics. In a similar vein, we claim that studies of human culture can benefit from mathematical and statistical approaches.

We are not alone in this view, of course, and digital humanities form an academic area that has recently emerged at the intersection between various traditions. There is, as yet, however, no satisfactory agreement as to what precisely digital humanities actually are and the term encompasses many things from digital archiving and text editing to data mining. Partly for this reason, digital humanities remain controversial. Approximately one century ago, physics split into theory and experiment and since then it split again and computational physics has emerged as a third branch. Whether something similar is happening in the humanities remains to be seen and whether or not quantitative approaches of the types described in the following chapters are categorised as digital humanities or something else is unclear. Either way, we have a coherent picture of what we mean by quantitative approaches to sources from the past. These are applications of, and development of, quantitative technologies to probe statistical information buried in folktales, fables, myths, legends, sagas, epics and histories. Digital humanities include useful activities to facilitate the harvesting of the required data, but our main focus herein is on the subsequent mathematical and statistical analyses.

Of the natural and mathematical sciences, those with a strong statistical bent (e.g., statistical physics, phylogenetics, linear algebra) offer navigable pathways into other disciplines. Statistical physics, for example, and the humanities have a common cousin in the social sciences. In the eighteenth century regularities were noticed in numbers of events such as births, marriages and deaths. This was surprising because at an individual level people are unpredictable. This partly motivated the development of statistical approaches to many-body physical systems by Boltzmann, Gibbs, Maxwell and others. In statistical physics, one deals with simplified models of systems comprising large numbers of identical constituents. Thanks to a remarkable property called universality, the large-scale structure of a many-particle critical system depends not on microscopic detail but only on a small number of global features such as the dimensionality of the system and whether the interactions between its constituents are short range or long range. Notions such as these—many-body systems, laws of averages and universality—are precisely why statistical physics offers such a good tool box for the social sciences. Add to that the power of computers to deal with many-body systems and big data sets and one realises why cross-disciplinary initiatives gathered pace in recent decades.

We have to acknowledge that mixing disciplines has not been uniformly welcomed down the years. Even in the nineteenth-century John Keats worried that the then new sciences of physics and chemistry might "unweave the rainbow", lamenting in his famous poem *Lamia*, that these rational methods could undermine the poetic beauty found in natural phenomena. What scientists see as beautiful might leave humanities experts aghast. Scientific models are usually idealised, which may prompt the accusation of over-simplification. Moreover, in the humanities, one does not have the luxury that every human in a society or every character in a body of literature is identical. Worse still, the idea of reducing a person or character to a simple number may be difficult to swallow. Nonetheless, experience and some of the following chapters show that, with a large number of people or characters or properties in the system, one can say something meaningful about the aggregate. To quote the statistician George E. P. Box: "Essentially, all models are wrong but some are useful."

Of course, we recognise that humanities scholars won't accept a number that is churned out by an algorithm as an end in itself or as a definitive answer. Instead, the quantitative approach may provide a heuristic device to discover patterns, much as they do in the natural sciences. These have then to be considered in the manner that is appropriate for the field and for the questions being addressed. In other words, quantitative tools may supply some answers but humanities provide the questions.

We begin our quantitative investigations into literature in the next chapter, where Robin Dunbar explores cognitive constraints on real life and literature. Dunbar demonstrates how stories have to strike a balance between being challenging, on the one hand, and being realistic enough to be interesting, on the other. He addresses the psychological underpinnings that make storytelling possible as well as explaining some of the functional benefits that derive from that activity. Dunbar's chapter shows how psychological theory can be combined with new statistical methods to deliver insights beyond those of more conventional literary approaches.

A few years ago, a new method was introduced to study and compare narrative texts. Instead of focusing on the literary or narrative bases of the texts, the approach, described in the chapter titled "A Networks Approach to Mythological Epics", involves extracting data for statistical analysis of character networks. It has clearly shown that social-network analysis forms a good bridge between different disciplines; it can connect science and humanities in joint research; it can depict old research questions in a new light and connect different phenomena belonging to the worlds of nature and culture. In this chapter, the networks approach is applied to epic narratives from four different European cultures of the past.

The chapter titled "Medieval Historical, Hagiographical and Biographical Networks" also presents comparative social-network analyses, but this time very much with a humanities lead. This chapter illustrates the differences between the perspectives, priorities and interests of humanities and natural sciences research. Whereas scientists gravitate to large data sets and look for universal properties, this contribution uses networks to examine the presentation of the same people and events in different sources through focussing on visualisation and local network properties. The combination of traditional qualitative and new quantitative approaches enables a new way of engaging with these sources, including an analysis of the importance of genre on the construction of social networks within texts.

Over the past two decades, Yuri Berezkin applied principal component analysis to his impressive catalogue of folklore and mythology motifs. As reported in the chapter "Peopling of the New World from Data on Distributions of Folklore Motifs", his research clearly shows that tales from Central and South America have stronger links to those from the Indo-Pacific belt of Asia than they have to mythology from North America. This suggests that the peopling of the Americas was a more diverse process than previously thought. Berezkin's suggestions have only very recently been confirmed in genetic studies. His database of 50,000 entries covering 2000 motifs from over 1000 traditions is currently available in Russian only. We consider it imperative that this vast resource be more widely available to scholars and researchers over the years to come and we hope that this book will help bring it the attention it deserves.

The chapter "Phylogenetics Meets Folklore: Bioinformatic Approaches to the Study of International Folktales" describes recent exciting applications of phylogenetics to folktales. Like the genes of biological organisms, traditional folktales mutate as they are transmitted from generation to generation. Jamshid Tehrani and Julien d'Huy are pioneers in the application of phylogenetic methods developed by evolutionary biologists to study the evolution of folktales. The extraordinary interest in this type of approach is indicated by the fact that Tehrani's paper on the phylogeny of *Little Red Riding Hood* was viewed over 73,000 times in a year and was one of the top 100 most discussed academic papers worldwide in the year it was published. Like Berezkin's analyses, the indications from Tehrani's and d'Huy's work are that oral tradition can preserve narrative motives over vast periods of time—many thousands of years. This level of robustness for non-physical entities is remarkable and has the potential to deliver a sort of "carbon-dating" for non-material culture.

The chapter titled "Analyses of a Virtual World" also presents a series of techniques that, in the future, could be applied to data extracted from ancient sources. In this chapter, Yurij Holovatch, Olesya Mryglod, Michael Szell, and Stefan Thurner analyse human behaviour in a virtual environment. Besides social structures, they are able to detect statistical similarities and differences between human-action dynamics in real and virtual worlds to analyse how society changes in time and how each element in that society acts correspondingly. The inter-event time distributions they observe signal a degree of "burstiness" in human dynamics. Inspired by the classification of life times of unstable elements in physics, they propose a way to quantify these distributions. They raise a number of interesting and important questions by allowing networks to form from different types of interactions, some positive and some negative. The importance of the preferential attachment scenario suggests that the more enemies a character has, the more likely they are to accrue more enemies. The same is not true for positive connections, however. Indeed, networks with positive connotations tend to be reciprocal, a property lacking in networks formed out of negative connotations. The chapter also illustrates the essential roles played by casual acquaintanceships in linking communities in virtual networks. Many methods developed can be applied to analyse epic and annalistic narratives, opening a vast area for future research.

In the chapter "Ghostscope: Conceptual Mapping of Supernatural Phenomena in a Large Folklore Corpus", Peter Broadwell and Tim Tangherlini consider the relationship between place and folklore. They present a method (called *GhostScope*) for representing the conceptual mapping of the environment by storytellers. *GhostScope* locates storytellers at a conceptual centre, relative to which places mentioned are mapped. They show how, in Denmark in this instance, folktales represent supernatural dangers from all angles and at all distances; giants, elves, robbers, mound dwellers, nisser and witches threaten from various locations and even from within. It would be fascinating to apply *GhostScope* to folktales from other countries to investigate common and differentiating features.

We switch to the investigations of vocabulary in the chapter called "Complex Networks of Words in Fables" where Yurij Holovatch and Vasyl Palchykov give an overview of the application of complex-network theory to the Ukrainian language used in two fables. The importance of this type of work in the current context is that it represents the start of an attempt to identify universal and specific features. The hope is that non-universal features may, in the future, help identify or even categorise genres of narrative. Such a programme is in its infancy and vastly more data are required. For example, would another set of children's literature deliver different results? Would epics or annals be different again?

We turn to the annals of Ireland in the chapter "Analysing and Restoring the Chronology of the Irish Annals" where Daniel McCarthy describes Ireland's extraordinary legacy of medieval chronicles. His interest is the chronological apparatus used to identify each year recorded in the annals. Centuries of scribal errors in the copying process led to historians dismissing the annals as hopelessly confused chronologically. Thanks to McCarthy's efforts, the correct chronologies have now been restored in many of these remarkable treasures.

The theme of medieval Irish chronicles is continued in the chapter titled "Mapping Literate Networks in Early Medieval Ireland", where Elva Johnston explores their wider potential use as data sets. Johnston explains how they are replete with geographical and personal details encoding a great deal of information. For example, the published genealogies alone contain information on more than 12,000 individuals who lived in Ireland before the twelfth century. Besides being attractive for character-network analyses, these contributions inspire the idea to use phylogenetics to track ancestry of errors in annals; chronicles with same errors may well have been copied from a single older source which may no longer be extant. Having the correct chronologies to hand, this also opens the way to apply some of the techniques developed in the chapter "Analyses of a Virtual World". Besides investigating static and dynamic network structures, it would be fascinating to study patterns in inter-event times, for example. In fact many of the approaches outlined in the previous chapters are applicable to the Irish chronicles (as well as those of other nations).

We return to the theme of comparative mythology in the chapter "How Quantitative Methods can Shed Light on a Problem of Comparative Mythology: The Myth of the Struggle for Supremacy Between Two Groups of Deities Reconsidered", where, following an overview of the oldest extant chronicle in Japan, David Weiß recalls historical interpretations of its account of struggles between two groups of deities. Pointing out that these fail to explain the numerous examples of comparable struggles in many of the world's other mythologies, Weiß recounts two major theories which seek to explain this. Each theory is based on the observation that the different mythologies considered have similar structures. Weiß's study is thus far classic comparative mythology, however, he goes on to suggest how some of the quantitative approaches of earlier chapters might enable us to test these hypotheses. While such quantitative methods will not (and should not) replace traditional approaches to comparative mythology, they might bring a certain degree of measurability to the field and help place it on a firmer fundament.

We hope that some of the above ideas and approaches will help to enhance old, but constantly fascinating fields of study—to complement, but certainly not to supplant, traditional techniques in the humanities. In terms of how data are prepared for quantitative analyses, many of the above examples are rather generic in their approaches but each has the potential to inspire new mathematical, statistical and algorithmic developments, aimed precisely at computational folkloristics and similar fields. These may in turn lead to exciting new insights and viewpoints and developments in applied mathematics and statistics.

Acknowledgements We would like to thank the Leverhulme Trust, the European Science Foundation and the European Union's FP7 programme for generous funding support. We also thank all the participants of the *Maths Meets Myths* workshops for helping develop the programme.

Cognitive and Network Constraints in Real Life and Literature

Robin Dunbar

Abstract Storytelling has played a major role in human evolution as a mechanism for engineering social cohesion. In large measure, this is because a shared worldview is an important basis for the formation not just of friendships but, more generally, of social communities. Storytelling thus provides the mechanism for the transmission of shared cultural icons and shared histories within a community. That being so, the effectiveness with which stories do their job is likely to be related to the storyteller's ability to make challenging yet realistic stories without overtaxing the listeners' abilities to comprehend the narrative. I summarise some of the constraints likely to act on this both in terms of community size and organisation and in terms of cognition, and explore their implications for storytelling.

Storytelling, in its many different forms, has played a central role in human evolution. Although we might distinguish, in this respect, fictional storytelling round the campfire from myths and religion, and from science (understanding the world), all of these involve someone offering an account of some virtual world that we do not directly experience. At some level or other, all of them have important advantages in that they allow individuals to acquire an understanding of how and why the world is as it is, and hence offer an important knowledge base for controlling their world. In addition, however, the first three provide particularly important benefits in terms of community cohesion. The stories that we know and love because we heard them so often at our mother's knee or, later, round the campfire, the myths that detail our origins as a community and why we share a common heritage, or the religious framework that we all sign up to—all mark us out as members of a particular community. Our shared understanding of why we are different from "that lot over the hill", our shared belief in a particular social or religious convention, identify us out as members of a specific community, and this

R. Dunbar (✉)
Department of Experimental Psychology, University of Oxford, South Parks Road, Oxford OX1 3UD, UK
e-mail: robin.dunbar@psy.ox.ac.uk

© Springer International Publishing Switzerland 2017
R. Kenna et al. (eds.), *Maths Meets Myths: Quantitative Approaches to Ancient Narratives*, Understanding Complex Systems,
DOI 10.1007/978-3-319-39445-9_2

helps to bond us as a group. Indeed, it turns out that these same features are the ones that determine the strength of our individual friendships (Curry and Dunbar 2013).

My main concern here, however, is not so much with the functional benefits that derive from storytelling in these many different forms, but rather with the psychological underpinnings that make them possible. Storytelling is possible at all only because we are willing to sit and listen to someone else (almost always in the singular) telling us something. We learn from that experience, and gradually build up a matrix of shared knowledge. That shared knowledge allows us to increase the efficiency of speech because we do not have to spell out everything on every occasion. Instead, we assume that we all understand the background, and this allows most of our conversations to consist of passing allusions and to be extraordinarily telegraphic as a result—full of circumlocutions and half-articulated assertions, unfinished sentences and obscure metaphors. I know that you can fill in the missing gaps because we share a common framework within which all this will make perfect sense—and complete nonsense to those who aren't members of our community, further allowing us to draw a clear line between us and them.

Our ability to understand story telling has, hitherto, been limited by lack of adequate tools to explore some of the underlying complexity in the psychological processes involved in storytelling. My claim will be that new developments in psychological theory and statistical analysis (notably network analysis methods) offer opportunities for novel insights beyond those that more conventional literary methods allow.

There are a number of important psychological constraints on what is possible in the context of storytelling. These have to do with the three separate features of our psychology. One is that our stories, or at least our fictional stories, should not be too far removed from real life. This can take a number of different forms, each of which is important in itself. The closer a story mirrors real life the more persuasive it is likely to be. We can tolerate some slippage, in that slight exaggerations of real life seem to be especially engaging, but the slippage cannot be too large or it all becomes implausible. This is essentially the idea of minimally counterintuitive concepts, originally developed in the cognitive science of religion: we are more likely to believe that our heroes can break the laws of physics (e.g. walk on water or pass through walls), but not by too much or it becomes implausible. And, of course, what our heroes can do we mere mortals cannot because we are subject to the full force of everyday physics. So things that are slight exaggerations push buttons for us, but there is a fine balance between the plausible and the implausible.

There is a second way in which the constraints of everyday reality are important, and this is in terms of how stories are structured, especially in respect of their network structure. If this deviates from the natural structure of everyday social networks, the story is likely to become incomprehensible because it overloads our ability to keep track of who is doing what with whom. Network methods may be able to provide us with novel insights into story structure because they reduce some of this natural complexity to simpler, more easily understood indices. Applications of network analysis methods to literature have been few and far between, but the approach has considerable potential because these methods have the capacity to provide insights into the structure of stories in ways that typically cannot be done by the more conventional methods of literary criticism.

The third class of cognitive constraint concerns the role of mentalising in our ability to create and understand exciting storylines. Mentalising, or mindreading, is the ability to understand what another individual believes about the world. The concept of mindreading has its origins in philosophy of mind, where the ability to understand one's own and others' intentions was seen as an important feature of the human condition (the so-called *intentional stance*: Dennett 1978). Formalised as "having a theory of mind" (or simply "theory of mind" for short), it identified the fact that I have a belief about your mindstate (your beliefs or intentions) as being central to many of our social interactions. These mindstates are associated with the use of intentional words—the class of words like "believe", "suppose", "intend", "wonder", and "think" that refer to the mind doing what we might think of as "serious work". Acquiring theory of mind (i.e. the ability to recognise that someone else has a belief about the world, and in particular one that may be different from my own belief) is something that happens quite early on in childhood (typically around the age of 5 years).

The capacity to mentalise is in fact recursive, potentially infinitely so (Dennett 1978), in the sense that A can *believe* that B *supposes* that C *wonders* whether D *intends* . . . and so on for as long as one might wish (with the individual intentionality terms italicised). Although this sequence could go on forever, in practice there appears to be a natural limit at five minds (including your own as the reader) imposed by the capacities of the human mind. Each of these belief states is defined as a level (or order) of intentionality. Understanding my own mind is equated with first order intentionality; understanding someone else's mindstate is second order; and the example I gave above is fifth order (with your mind as reader as the anchor: when you read it, you *realise* that A *believes* . . . .etc). Not only does this place a limit on the complexity of the sentences we can construct, but also, seemingly, on the number of different individuals one can have in a conversation.

I will consider each of these approaches in turn, and, on the basis of the examples I offer, try to provide some evidence that investing more heavily in all three will pay dividends.

Reality-Mining in Literature

Let me first illustrate how reality intrudes into storytelling, at least when those stories purport to tell us about events in the real world. Purely mythical stories are usually prefaced with some cues that they concern people or events that happen outside the confines of everyday experience. In contrast, stories that that are about everyday events or even real historical events must abide by the conventional experiences of that realm. The Icelandic family sagas, mostly composed in the late twelfth and early thirteenth centuries, detail genuine historical events, but, as is inevitable with early historical documents, with a variable degree of elaboration and exaggeration. The opportunity for exaggeration notwithstanding, the authors of these histories write with remarkable consistency in ensuring that their stories fit with what we would expect from real life. Hamilton's (1964) theory of kin selection, for example, is one of the central pillars of modern evolutionary theory and states that individuals will behave in such as way as to favour their relatives over unrelated individuals, unless the benefits to the actor of doing so favour the unrelated individual. In an analysis of two major sagas, *Njalssaga* and the *Orkneyingasaga*, we found that the murders reported in the sagas followed this pattern: when the benefits of the murder were low (following a spontaneous fight after an insult or a drunken brawl) close relatives were much less likely to be murdered than unrelated individuals, but when the benefit from the death was significant (acquisition of land, or inheriting an earlship) close relatives were not spared (Fig. 1).

Kinship is important in human affairs because kin are the one group of people who will come to one's aid come what may. This is a simple consequence of Hamilton's theory of kin selection: evolution drives a willingness to behave altruistically towards kin in a way that does not happen towards non-kin (a claim supported by considerable empirical evidence: see, for example, Madsen et al. 2007 and references therein). Kin provide an important source of support in dangerous contexts, such as when one is under attack or wants to attack others. One example of that is provided by evidence from the sagas that individuals were willing to commit murder only when they had more kin available to come to their support than their victims did (Fig. 2). They were seemingly not prepared to risk murdering someone if that individual had a large number of relatives in Iceland, because relatives were entitled to pursue revenge strategies or make claims against the murderer.

These twoexamples provide some evidence that storytellers cannot stretch the bounds of real world plausibility, even if that would suit the purposes of the story

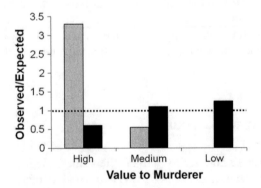

Fig. 1 Relative frequency with which individuals murdered others as a function of whether they were kin (related as paternal cousins or closer) (*hatched bars*) or not (*black bars*) and the benefit obtained by the murderer. Relative frequency is the observed number of murders divided by the number one would expect if murderers were distributed at random across the population. High benefit includes acquisition of land or titles; low benefit refers to murders arising from trivial altercations like insults or drunken brawls. Redrawn from Dunbar et al. (1995a)

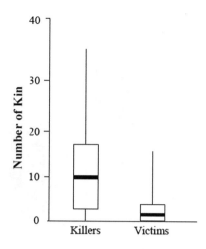

Fig. 2 Number of kin for murderers and their victims in three Icelandic sagas (*Njal's Saga*, *Egil's Saga* and the *Laxdaela Saga*). The *thick black line* gives the median; the *box* gives the distribution of the 50 % of values closest to the median; and the *whiskers* the distribution of 95 % of the data. *Source*: Palmstierna et al. (2016)

they want to tell. They have to stick within what is plausible and reasonable if they are to carry their audience with them.

Natural Network Structure

Human social networks have a very distinctive structure. If you rank all the members of your network in a linear sequence based on the emotional intensity of your relationship with them, they naturally fall into distinct sets that comprise, successively, 5, 10, 35 and 100 individuals. When these are progressively added together, they comprise a natural series that has a very distinct scaling ratio of three: in other words, each circle is about three times larger than the circle inside it (5, 15, 50, 150). This pattern reappears in many different social contexts, ranging from personal offline (Hill and Dunbar 2003; Zhou et al. 2005) and online social networks (Passarella et al. 2012; Arnabaldi et al. 2013) and the natural organisation of hunter gatherer societies (Hamilton et al. 2007) and the community structure of online gaming worlds (Fuchs et al. 2014).[1] In part, this distinctive structure is created by how we divide our available time among the various members of our social networks (Sutcliffe et al. 2012a, b). These structural constraints have important implications for the way personal social networks are structured, since they place limits on, for example, indices like the connectivity of the individuals in a network.

Another important constraint on the size of natural human groups is the limit on conversation group sizes. These turn out to have a very consistent upper limit at four individuals (Dunbar et al. 1995b; Dezecache and Dunbar 2012). It seems that we

---

[1] Editors' note: See, also, the Chapter "Analyses of a Virtual World" by Holovatch, Mryglod, Szell and Thurner.

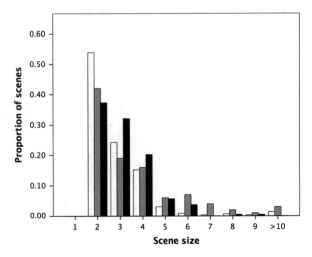

Fig. 3 Proportion of scenes of different size in 12 'hyperlink' genre films (*white bars*), 10 Shakespeare's plays (*grey bars*) and natural conversations (*black bars*). *Sources*: Krems and Dunbar (2013), Stiller et al. (2004) and Dezecache and Dunbar (2012), respectively

are unable to maintain a coherent conversation with more than four people for any length of time: a conversation that acquires a fifth person very quickly breaks up into two sub-conversations, often with constantly revolving membership as individuals switch from one conversation to another. It is not clear what imposes this limit, but it may have to do with the number of different mindstates we can keep track of at any one time (Krems et al. 2016).

Figure 3 plots the distribution of conversation sizes in 193 natural conversations, 10 of Shakespeare's plays and 12 'hyperlink' genre films. (Hyperlink films are a genre of films that seek to break through the natural limits on communication such that individuals' behaviour can influence each other even though they are in different times and places. They include films like *Crash*, *Babel*, *Love Actually*, *Traffic*, etc.) For the plays and films, a conversation is defined as a scene, with a new 'scene' starting whenever a character left or joined an existing conversation. These three very different kinds of data have virtually identical patterns. In all three cases, 85 % of conversations contained four or fewer individuals (or characters), and 90 % contain five or fewer. It seems that neither Shakespeare nor modern scriptwriters are willing to break the patterns seen in natural conversations. I suggest that this is because, if even if they could do it themselves, they would rapidly lose their audiences.

Figure 4 plots the clustering coefficient for the hyperlink film sample, along with the mean values for the 10 Shakespeare plays, a set of 10 contemporary romantic genre films (examples include *A League of Their Own*, *First Wives Club*, *Pride and Prejudice*, *Sex and the City* and *The Sisterhood of the Traveling Pants*), and the plays of the Russian playwright Anton Chekov and the Irish playwright George Bernard Shaw. The clustering coefficient is the probability that if A is linked to B and B

Fig. 4 Frequency of
clustering coefficients in 12
individual hyperlink films
(*bars*). Also plotted on the
graph are the mean values for
10 Shakespeare plays (*solid
vertical line*), 10 romantic
genre films (*dashed line*),
Chekov's plays (*fine dots*)
and the plays of George
Bernard Shaw (*spaced dots*).
The *arrow* marks the
equivalent frequency in real
life networks. *Source*: Krems
and Dunbar (2013)

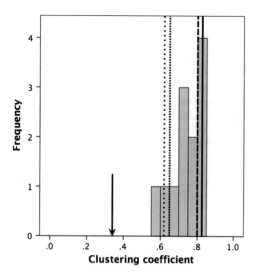

is linked to C (i.e. have appeared together in the same conversation/scene), then A
and C will also be linked. In real world networks, this kind of triad, in which all
three connections are positive, are particularly common, and account for about a
third of all relationship triads in both real life (Leskovec et al. 2010) and online
gaming environments (Szell et al. 2010). Note the high proportion of such triads in
the drama samples (far higher in fact than in normal everyday networks, demarcated
by the arrow), and the broad similarity between the five different genre of drama.
Nonetheless, both Chekov and Shaw have relatively low values compared to the
other three, though still well above the norm for everyday networks. This suggests
that these two nineteenth century dramatists typically used a more diffuse network
structure than did either Shakespeare or contemporary film scriptwriters. Whether
this reflects the particular subject matter of their plays, or the structural design they
imposed on the action in their plays, remains to be determined.

This broad similarity does not, however, extend to all network indices. Figure 5
plots the density of ties (the number of direct connections that each character has)
in these same films and plays. Hyperlink films clearly differ strikingly from the
other genre in having a much lower proportion of directly connected characters
(characters that appear together in conversations), despite the fact that they don't
differ that much in total cast size from, at least, the Shakespeare plays. Chekov and
Shaw plays have a particularly high density, mainly because they have relatively
small cast lists compared to the other genres.

The effect of cast size on density is illustrated in Fig. 6a for the Shakespeare
plays. Essentially, density declines as the number of characters increases (Pearson
correlation $r = -0.877$, $p = 0.001$, indicating a very strong negative relationship).
In part, of course, this is simply a consequence of the fact that 'conversation' group
size and degree (the number of individuals each character is directly connected to)
remain constant despite increases in cast size. There is some suggestion that this

Fig. 5 Frequency
distribution of the density
(proportion of connected
nodes) in hyperlink genre
films. Also plotted on the
graph are the mean values for
10 Shakespeare plays (*solid
line*), 10 romantic genre films
(*dashed line*), Chekov's plays
(*fine dots*) and the plays of
George Bernard Shaw (*long
dots*). Note the log scale of
the X-axis. *Source*: Krems
and Dunbar (2013)

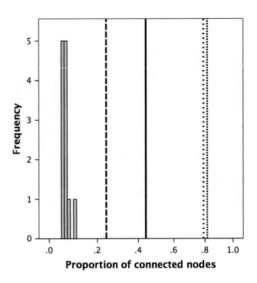

index reaches a lower asymptote at a value of around 0.2 when cast size exceeds 45. Figure 6b plots the average path length (the number of connections that separate any two characters in the play) for these same plays. Average path length rises with cast size ($r = 0.918$, $p < 0.001$), with a possible tendency towards an asymptotic value somewhere around 2.0 when cast size is 50 or so. In contrast, the clustering coefficient (Fig. 6c) seems to behave rather differently. While there is a weak, non-significant negative relationship with cast size (the bigger the cast list, the less likely triads are to be fully connected: $r = -0.499$, $p = 0.124$), the distribution seems to have a rather obvious upper bound (denoted by the dashed line). In other words, there is an upper limit beyond which it is not possible to go, but any value below that limit is possible. That limit is independent of cast size and the three classes of plays do not differ in this respect. In fact, the regression line for this upper bound has a standardised slope coefficient of $b = -0.981$, which is close enough to $b = -1$ to suggest that the upper limit on clustering coefficient is a constant function of cast size. This likely suggests a constraint imposed by the limit on conversation group size and the typical time frame of plays: you can only include so many conversations in a play of a given length (in this case, about 2.5 h) and so some triads just have to be left to the audience's imagination.

The distributions in Fig. 6a, b could be interpreted as being asymptotic, but another interpretation is possible. Relative to the pattern observed in the other plays, our one example of a history (*Richard III*) has a disproportionately low frequency of connected dyads and a rather higher mean path length (i.e. relatively longer path lengths between dyads) than one might expect. It is not clear whether this is a feature of histories as a genre or simply due to the natural behaviour of networks when their size gets large (irrespective of genre)—or just an oddity of this particular play, which is known to be difficult to stage because of its large cast. One potential problem that a playwright faces with histories is that they have less control over the number of

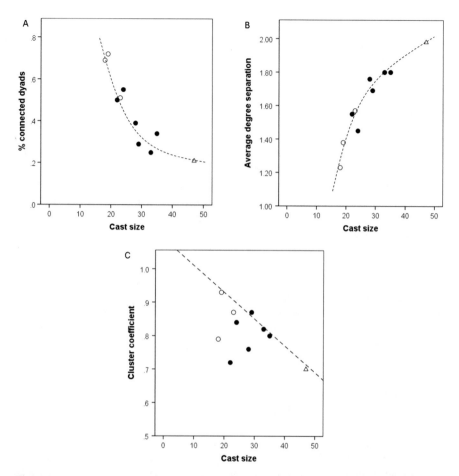

Fig. 6 For each of the 10 Shakespeare plays, differentiated between tragedies (*solid symbols*), comedies (*open symbols*) and the one history in the sample (*Richard III*) (*triangle*), I plot (**a**) density (proportion of all dyads that are actually connected, i.e. appear in a conversation together), (**b**) average path length (the mean number of links required to get from any one character to any other) and (**c**) the clustering coefficient (the proportion of all triads in which all three members are connected). In (**c**), the data suggest that there is an upper bound (shown by the *dashed line*) beyond which values cannot go. *Source*: Stiller et al. (2004)

characters and their interactions: the actual events may necessitate the inclusion of a larger number of characters than can is ideal for the audience to handle, and in real life they may have interacted more than would be desirable for a structurally simple script. *Richard III* is commonly viewed as difficult to stage, and directors often combine characters to reduce some of the complexity. In terms of size, its cast of 47 primary characters is certainly pushing the limit on one of the natural grouping layers of human social networks (the 50-layer). If the relationships in Fig. 6a, b are actually linear, a cast of 47 ought to have a density that is close to 0 and an average path length of about 2. Since being in a fully connected triad without ever actually meeting in a scene would be rather challenging, some twisting of the plot structure might be necessary to accommodate so many characters.

Mentalising Constraints

Mentalising, the capacity to understand the mindstates of other individuals, is one of the core cognitive abilities that distinguish humans from other animals. In the present context, mentalising, or mindreading as it is sometimes known, is what allows us to engage in a fictional world, to imagine a world that is distinct from the world we actually inhabit—one where we can engage in the activities and even mindstates of fictional characters within that virtual world. Without this capacity, we would be limited to a literalist interface with the world we live in, much as is the case for all other animals. This ability to step back from the world we experience directly and imagine a virtual one depends on our being able to separate reality from 'virtuality'. Just how important this is is highlighted by the case of autistic individuals, who lack theory of mind (the first step on the recursive process of mentalising). Without theory of mind (second order intentionality), autists take the world exactly as they find it and have difficulty imagining alternatives, or even imagining that the world could have been different to the way it appears to be. As a result, they have difficulty understanding metaphors, and instead take the meanings of sentences to be exactly what the words they contain literally mean (Happé 1994). While much work has been done on theory of mind, in fact this is a competence that children acquire as early as 5 years of age. Adults can do significantly better than 5-year-olds, and we have shown in a number of studies (Kinderman et al. 1998; Stiller and Dunbar 2007) that the typical upper limit for normal adults is fifth order intentionality, with only modest and diminishing numbers of people being able to manage sixth order or better (Fig. 7).

If we consider the mental work that someone has to do at a play, the magnitude of the task becomes clear. Think about what is involved when we watch Shakespeare's *Othello*. Here, at least as far as the plot of the play is concerned, we have to believe that Iago *intends* that Othello *believes* that Desdemona *loves* Cassio, and in the limit that her love is reciprocated by Cassio (i.e. Cassio also *loves* Desdemona). What makes the difference here is the last intentional state (Cassio's reciprocating Desdemona's love). Without that, we have a fairly boring narrative story: Desdemona has the 'hots' for Cassio, but in the absence of any action on Cassio's part this isn't very exciting and Othello shouldn't have cared less. What makes Shakespeare's story grip us is that Othello is led to believe that Desdemona's mindstate (being in love with Cassio) is reciprocated by Cassio, because now there is every chance that they might actually run away together (thus publicly embarrassing Othello, never mind causing him emotional pain) or, worse still, arrange for Othello to be bumped off so that Cassio could take his place as the General of the Venetian army.

The significance of this is that the audience has to handle four separate mindstates simultaneously, and, since they can only do that by having a mindstate of their own to do this with, this means that they have to work at their natural limit at fifth order intentionality. In other words, in a play like *Othello* with four main protagonists, the typical audience member is being pushed to their cognitive limits in terms of mentalising. But to be able to do that, the dramatist (in this case, Shakespeare) has to be able to work at one level higher (sixth order). In other words, while 65 % of the population will be able to appreciate the play because they can work at the requisite intentionality level, only about 35 % of the population have the cognitive ability (at least in terms of mentalising) to actually write the play.

One obvious prediction from this might be that an audience member's ability to enjoy a play (or, equivalently, a reader's ability to enjoy a novel) is determined by how far the author can push the audience. In other words, for someone with fifth order mentalising competence, a story that requires fourth or fifth order mentalising on their part is likely to be rated as more engaging or exciting than one that requires only second or third order mentalising, whereas a story that requires sixth or seventh order mentalising is likely to be seen as difficult and less enjoyable. Similarly, someone with third order mentalising competences might enjoy a third order story, but find fourth, fifth and sixth order stories increasingly more difficult to get to grips with, and hence less enjoyable. Carney et al. (2014) tested this using short (1000-word) story vignettes from two different genres (a spy story and a story about romantic relationships) written at third or fifth order intentionality. Subjects found fifth order spy stories less engaging than third order spy stories, but they rated fifth order romantic stories as more enjoyable than third order ones. Thus, familiarity (or, alternatively, closeness to everyday experience) may influence our ability to follow the twists and turns in a plot. In similar vein, subjects rating jokes told by

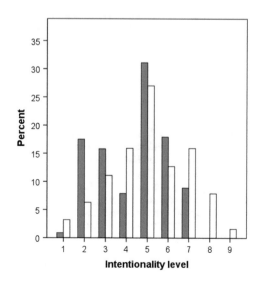

Fig. 7 Distribution of mentalising competences (indexed as the highest level at which individuals successfully passed these tasks) in two different studies: Kinderman et al. (1998) (*grey bars*) and Stiller and Dunbar (2007) (*white bars*). Kinderman et al. only tested subjects up to sixth order intentionality, and a proportion of these subjects would have passed at higher levels; for illustrative purposes, those passing at this level have been partitioned in a 2:1 ratio between 6th and 7th orders

famous comedians found higher mentalising order jokes more enjoyable than less demanding jokes (Launay and Dunbar 2016).

In composing his plays, Shakespeare seems to have paid very close attention to the demands he was placing on his audiences. In an analysis of the mentalising states involved in the same 10 Shakespeare plays and a sample of natural conversations on a University campus, Krems et al. (2016) found that the number of people involved in a conversation (hence the number of intentionality levels involved) was one whole mindstate level lower when the conversation involved an absent party's mindstate than when it involved either factual events offstage or the mindstate of one of the people involved on stage in the conversation. What is quite remarkable is how closely Shakespeare followed the natural rhythms of conversation, adjusting naturally for features that are really quite subtle.

Conclusions

I have offered three ways in which we might gain novel insights into storytelling in its many different forms by exploiting powerful theoretical frameworks on offer in psychology and evolutionary biology, combined with novel statistical methods like network analysis. These have so far been relatively underused in the humanities, but their potential to offer insights is considerable when handled with appropriate care and informed application. Here, I have done no more than point the way with some simple, but hopefully informative, examples. These examples should not be taken as indicating the limits of such applications. Rather, they are intended to provide encouragement and point the way ahead.

Acknowledgements My research is funded by a European research Council Advanced Investigator grant. This article also forms part of the output of the Calleva Research Centre, Magdalen College, Oxford.

References

Arnabaldi, V., Conti, M., Passarella, A., & Dunbar, R. I. M. (2013). Dynamics of personal social relationships in online social networks: A study on Twitter. *COSN'13 Proceedings of the First ACM Conference on Online Social Networks*, pp. 15–26.
Carney, J., Wlodarski, R., & Dunbar, R. I. M. (2014). Inference or enaction? The influence of genre on the narrative processing of other minds. *PLoS One, 9*, e114172.
Curry, O., & Dunbar, R. I. M. (2013). Do birds of a feather flock together? The relationship between similarity and altruism in social networks. *Human Nature, 24*, 336–347.
Dennett, D. (1978). Beliefs about beliefs. *Behavioral and Brain Sciences, 1*, 568–570.
Dezecache, G., & Dunbar, R. I. M. (2012). Sharing the joke: The size of natural laughter groups. *Evolution and Human Behavior, 33*, 775–779.
Dunbar, R. I. M., Clark, A., & Hurst, N. L. (1995a). Conflict and cooperation among the vikings: Contingent behavioural decisions. *Ethology and Sociobiology, 16*, 233–246.

Dunbar, R. I. M., Duncan, N. B., & Nettle, D. (1995b). Size and structure of freely forming conversational groups. *Human Nature, 6*, 67–78.

Fuchs, B., Sornette, D., & Thurner, S. (2014). Fractal multi-level organisation of human groups in a virtual world. *Scientific Reports, 4*, 6526.

Hamilton, M. J., Milne, B. T., Walker, R. S., Burger, O., & Brown, J. H. (2007). The complex structure of hunter-gatherer social networks. *Proceedings of the Royal Society London, 274B*, 2195–2202.

Happé, F. (1994). *Autism: An introduction to psychological theory*. London: University College London Press.

Hill, R. A., & Dunbar, R. I. M. (2003). Social network size in humans. *Human Nature, 14*, 53–72.

Kinderman, P., Dunbar, R. I. M., & Bentall, R. P. (1998). Theory-of-mind deficits and causal attributions. *British Journal of Psychology, 89*, 191–204.

Krems, J. A., & Dunbar, R. I. M. (2013). Clique size and network characteristics in hyperlink cinema: Constraints of evolved psychology. *Human Nature, 24*, 414–429.

Krems, J. A., Dunbar, R. I. M., & Neuberg, S. L. (2016). Something to talk about: Conversations are constrained by mentalising abilities. *Evolution and Human Behavior* (in press).

Launay, J., & Dunbar, R. I. M. (2016). The complexity of jokes is determined by cognitive constraints on mentalising. *Human Nature, 27*, 130–140.

Leskovec, J., Huttenlocher, D., & Kleinberg, J. (2010). Signed networks in social media. *ACM SIGCHI Conference on Human Factors in Computing Systems (CHI)*, 2010.

Madsen, E., Tunney, R., Fieldman, G., Plotkin, H., Dunbar, R. I. M., Richardson, J., & McFarland, D. (2007). Kinship and altruism: A cross-cultural experimental study. *British Journal of Psychology, 98*, 339–359.

Palmstierna, M., Frangou, A., Wallette, A., & Dunbar, R. I. M. (2016). Family counts: Deciding when to murder among the Icelandic Vikings. (submitted)

Passarella, A., Dunbar, R. I. M., Conti, M., & Pezzoni, F. (2012). Ego network models for future internet social networking environments. *Computer Communications, 35*, 2201–2217.

Stiller, J., & Dunbar, R. I. M. (2007). Perspective-taking and memory capacity predict social network size. *Social Networks, 29*, 93–104.

Stiller, J., Nettle, D., & Dunbar, R. I. M. (2004). The small world of Shakespeare's plays. *Human Nature, 14*, 397–408.

Sutcliffe, A. J., Dunbar, R. I. M. Binder, J., & Arrow, H. (2012a). Relationships and the social brain: Integrating psychological and evolutionary perspectives. *British Journal of Psychology, 103*, 149–168.

Sutcliffe, A. J., Wang, D., & Dunbar, R. I. M. (2012b). Social relationships and the emergence of social networks. *Journal of Artificial Societies and Social Simulation, 15*, 3.

Szell, M., Lambiotte, R., & Thurner, S. (2010). Multirelational organization of large-scale social networks in an online world. *Proceedings of the National Academy of Sciences USA, 107*, 13636–13641.

Zhou, W.-X., Sornette, D., Hill, R. A., & Dunbar, R. I. M. (2005). Discrete hierarchical organization of social group sizes. *Proceedings of the Royal Society London, 272B*, 439–444.

A Networks Approach to Mythological Epics

Ralph Kenna and Pádraig MacCarron

Abstract In recent years, a new approach to the analysis of ancient texts and narratives has been developed. The method draws on network science, the study of systems with interacting elements that can be represented mathematically by graphs. Network science itself is associated with statistical physics, a branch of physics which employs probability theory and statistics to capture properties of large collections of interacting entities. Many ancient narratives—from the sagas of Icelanders, through epics such as the *Iliad* and *Beowulf*, to the stories contained in medieval Irish manuscripts—record multiple interactions between sometimes vast numbers of characters and social network analysis is an excellent tool to quantify their collective properties. By capturing the interconnectedness of their underlying social structures, such narratives can be compared to each other and to other genres of literature, past and present, as well as to modern-day social networks. Here we review the main ideas behind this new approach to comparative mythology and the interrelationships of characters appearing in epic narratives. We demonstrate that, by quantitatively comparing structural properties of ancient narratives, this new approach to the humanities can deliver new comparisons, observations and insights.

Introduction

In recent times, complex networks have emerged as a popular field of study with very broad applicability. In 2012, it was applied for the first time to a comparative study of three mythological epics: the Greek *Iliad*, the Anglo-Saxon *Beowulf*, and the Irish *Táin Bó Cuailnge* (MacCarron and Kenna 2012). That first paper generated

R. Kenna (✉)
Applied Mathematics Research Centre, Coventry University, CV1 5FB Coventry, UK

The Doctoral College for the Statistical Physics of Complex Systems,
Leipzig-Lorraine-Lviv-Coventry (L⁴), 04009 Leipzig, Germany
e-mail: r.kenna@coventry.ac.uk

P. MacCarron
Department of Experimental Psychology, University of Oxford, South Parks Road, Oxford OX1 3UD, UK

© Springer International Publishing Switzerland 2017
R. Kenna et al. (eds.), *Maths Meets Myths: Quantitative Approaches to Ancient Narratives*, Understanding Complex Systems,
DOI 10.1007/978-3-319-39445-9_3

an enormous amount of interest both within academia and amongst the public at large, not least because of its uncommonly high interdisciplinary nature. Since then, a number of other epic narratives from the ancient past have been analysed using network science, including the Icelandic sagas (*Íslendinga Sögur*) in another paper which captured the imaginations of academics and the public alike (MacCarron and Kenna 2013). Here we summarise some of these investigations and their results to illustrate of the potential power of applying maths to myths.

We begin by contextualising network sciences and graph theory in general— where it came from, its long history within mathematics, computer science and separately in sociology. We explain why it burst onto the scene in the 1990s as an exciting interdisciplinary field which gave potential for multiple collaborations across very different disciplines. We then move on to review elements of the very recent applications of network science to ancient narratives and show that it is an ideal approach to compare and contrast within and between genres. We look specifically at the *Iliad*, *Beowulf*, the *Táin Bó Cuailnge* and *Njáls saga*, the latter being the one of the most extensive of the *Íslendinga sögur*. The main questions we wish to address are as follows: (1) which network features are common to the societies described in the four stories? (2) what properties differ between the societies and thus allow us to quantitatively distinguish one from another? Answers to the first question may help us identify features which are universal to all tales or at least to tales within a given genre. The second question may open up new ways to compare between genres and narratives. By comparing to documented properties of real social networks, these may also help us to decide whether the society described in one narrative is more realistic than that of another. This, in turn, combined with information from other disciplines, may help inform as to the potential historicity underlying some aspects of the ancient texts.

Mythologies form foundation stones for a multitude of human cultures and are amongst the brightest gems of our shared cultural inheritance. They have persisted from before recorded history and still fascinate us in the present day. As we attempt to build a system to facilitate quantitative intercultural comparisons, we position narratives of various genres, cultures and epochs in a multidimensional metric space. Thus far, only a very small number of the world's ancient narratives have been mapped out in the form of networks and this can be considered to be the start of a programme of work. It is desirable that, in the long term, much of the world's complex literature be charted and analysed using complex networks. This will provide a new, quantitative way to compare and contrast bygone and current cultures through distances between them in the space of their mythologies.

Network Science: A New Tool in Comparative Mythology

While complex networks have become very popular in recent years, the subject actually has a long history in mathematics. Graph theory can be traced back to a problem tackled by Leonhard Euler in 1736. The question he addressed was whether one could find a path through the city of Königsberg (now Kaliningrad, in the Russian exclave between Poland and Lithuania) which traversed each of its seven bridges once only. In this problem the land masses are considered to be nodes or vertices of a network and the bridges are links or edges between them. Euler proved that the sought-after path does not exist and his study laid the foundations for graph theory. Another early example of graph theory, still commonly taught in computer science, is the problem of the *knight's tour*. This involves a sequence of moves by a knight on a chessboard wherein a square can only be visited once. The earliest example of graph theory applied to sociology is usually attributed to Jacob Moreno (Moreno 1934), who produced graphical representations of social links between individuals called sociograms. These are now more commonly referred to as social networks. Early graphs and sociograms were usually small enough to perform exact mathematical calculations or to represent them visually on paper. They typically had only a few vertices—at most a few dozen e.g., one can use sociograms to ask what happens to the overall connectivity of a network if a certain individual is removed, in an attempt to determine which individuals in a network have the most influence.

In recent years, vast online resources such as an internet movie database, electronic journals, as well as online social networks, allow us to gather data on a much larger scale than before. For enormous, highly connected networks, one is less interested in the effect of removing a single vertex—for such a move is likely to deliver only insignificant change. Instead, one now has to consider statistical questions such as the likely effects of removing a certain percentage of nodes (Newman 2006). The change in focus to large scale networks was accompanied by the development of methods to analyse their structures. Complex network theory has since been extended and applied to the study of a large range of areas, such as the internet (Cheswick and Burch 1999), epidemiology (Moore and Newman 2000), polymers (Scala et al. 2001), computer science (Myers 2003), food-webs (Dunne et al. 2004), astrophysics (Paczuski and Hughes 2004), economics (Schweitzer et al. 2009), power grids (Buldyrev et al. 2010), transport networks (von Ferber et al. 2012), linguistics (Holovatch and Palchykov 2007 and the chapter in this volume titled "Complex Networks of Words in Fables") and more (Albert and Barabási 2002). Thus the application of network science to humanities was a natural next step.

Many of the tools used in the study of complex networks are inspired by statistical physics. A key concept in that discipline is the notion of *universality* of critical phenomena. This is the idea that, despite their differences at a microscopic level, the macroscopic properties of systems undergoing phase transitions (so-called critical systems) depend only upon a few parameters such as symmetries and dimensions. Critical systems can therefore be categorised according to which universality class they belong to and these, in turn, are classified by a small set of numbers called critical exponents. There is a similar, albeit qualitative, concept in comparative mythology. In 1949, Joseph Campbell described the concept of the monomyth as a "template" supposedly common to a broad range of mythological tales (Campbell 1949). Critics have since argued that the monomyth is too broad a concept to be of practical use. Nonetheless, the similarity to the notion of universality in the statistical physics of critical phenomena inspired an attempt to quantitatively categorise epic narratives according to their network properties. These can also be encapsulated in a small set of numbers which aids comparison between them.

In a first application to comparative mythology, the networks underlying three iconic mythological narratives were analysed with a view to identifying common and distinguishing quantitative features (MacCarron and Kenna 2012). Of the three narratives, the *Iliad*, a Greek classic, and *Beowulf*, an Anglo-Saxon text, are mostly believed by antiquarians to be partly historically based while the third, the Irish epic *Táin Bó Cuailnge*, is often considered to be fictional. The original network analysis involved an attempt to discriminate real from imaginary elements of social networks and place mythological narratives on the spectrum between them. The scales of these networks are between the two extremes mentioned above—involving about 70–700 vertices. This allows both approaches to be used—a statistical one, on the one hand, to determine the macroscopic or global properties of the social networks described in the texts and a microscopic or local one to understand the influence of individual characters. By comparing the *Táin Bó Cuailnge* to the *Iliad* and *Beowulf* it was suggested that the apparent artificiality of the social network underlying the Irish narrative can be traced back to anomalous features associated with only six characters. Speculating that these might be amalgams of several entities or proxies, it was suggested that the plausibility of the Irish societal structure might be comparable to the others from a network-theoretic point of view.

Since then, a number of other epic narratives from the ancient past have been analysed using network science, including the *Íslendinga sögur* (MacCarron and Kenna 2013). The individual saga networks analysed, such as that of *Njáls saga*, have many properties of real social networks and comparison to the three other European tales reveals the relative importance of conflict in the narratives. In the *Iliad*, *Beowulf*, and the *Táin Bó Cuailnge*, conflict is an important element in that hostile links are generally formed when characters who were not acquainted meet on the battlefield. This is quite different for the *Íslendinga sögur*, for which many hostile links are due to blood feuds as opposed to regions at war. There, hostile links are often formed between characters that are already acquainted. For this reason, and quite unlike in the other three, there is little difference between the properties of the *Njáls saga* network and the positive subnetwork formed by omitting hostile interactions. If a comparison had to be drawn, however, *Njáls saga* is most similar to the positive *Iliad* subnetwork.

To understand how these observations were arrived at we next outline some of the essentials of complex networks. We then explain how it was applied to aforementioned narratives before moving on to suggest how the approach may be used in the future.

Elements of Complex-Network Analysis

A network is a set of vertices which are connected by edges. In a social network, the vertices are representations of people and the edges represent interactions or acquaintanceships between them. Here we are interested in *character networks*,

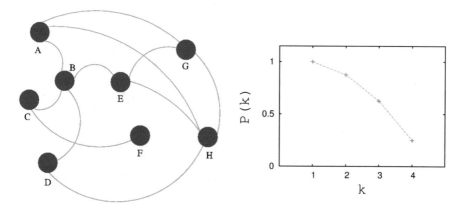

Fig. 1 An example of a network and its degree distribution. In the *left panel*, the undirected lines linking nodes (A, B, C,...) are edges of the network. In the *right panel*, $P(k)$ represents the cumulative degree frequency (percentage of nodes with degree k or more)

wherein the links are between characters appearing in a given text. A number of statistical tools have been introduced to describe and quantify various properties of networks. Here we briefly describe these tools with minimum mathematics. The reader interested in the mathematical details is referred to the literature such as Newman (2010) and Estrada (2011) and references therein.

The *degree* of a node in a network is the number of edges it has and we represent it by the variable k. In a social network the degree is the number of people linked to an individual. In the narratives we shall deal with, the degree of a character is the number of other characters he or she interacts with e.g., in the left panel of Fig. 1, the degrees of nodes A, B and C are respectively 3, 4 and 2. The degree varies from vertex to vertex and the spread of different degrees across the vertices of a network is called the *degree distribution*. To represent this, we often use the complementary cumulative probability distribution function $P(k)$. This is the probability that a given vertex has degree k or above. For example, in Fig. 1, 100 % of nodes have degree $k = 1$ or more by default. For that network the proportion of vertices with degrees k greater than or equal to 2, 3 and 4 are 7/8, 5/8 and 1/4, respectively. Representing these as $P(k)$, and plotting versus k, results in the degree distribution pictured on the right of Fig. 1.

Complex networks tend to have right-skewed degree distributions (Newman 2010). These heavy tails are frequently due to a small number of vertices having very large degrees. Such hubs tend to have a disproportionate amount of connections and can strongly influence the properties of networks. A common behaviour, often used as a first approximation fitted to degree distributions of social networks, is the power law (Albert and Barabási 2002). This is expressed as $p(k) \sim k^{-\gamma}$ where the symbol "$\sim$" means "behaves as". Such a formula lacks a scale because, if we multiply k by 2, say, this simply results in multiplying $p(k)$ by a constant so that $p(2k) \sim k^{-\gamma}$ again. This means that $P(2k)$ has the same shape as $P(k)$. Such networks are called *scale free*. The exponent γ, in these cases, is often found to be between 2 and 3, meaning that only few people tend to interact with a very large number of other people. A similar right-skewed function, known as the truncated power law $p(k) \sim k^{-\gamma} \exp(-k/\kappa)$, contains a power-law regime followed by a sharp cut-off where the tail decays much faster. These distributions are not scale free because the location of the cut-off κ sets a scale. In these networks the highest connected nodes are not as dominant as in the pure power-law.

In Fig. 1, the path length between nodes A and F is 3 because the shortest chain (A $\rightarrow$ B $\rightarrow$ C $\rightarrow$ F) has three edges. There is a well-known notion in sociology called *six degrees of separation*. (Here the word "degree" has a different meaning to that introduced above as k.) This is related to the idea that, despite the world's population of over seven billion, everyone is, on average, only about six steps away from any other person in the sense that a chain of six acquaintanceships or fewer can connect any two people. In this case the number six is an example of the mathematical notion of *average path length*. The average path length is a measure of the *connectivity* of the network and we denote it by ℓ. It tends to be very short in social networks (as the number six is small compared to the world's population).

Another measure of connectivity is the *clustering coefficient*, which gives an indication of how cliqued a network is. If a node is connected to other nodes, then the clustering coefficient gives the probability that these neighbours are also connected to each other. To calculate it for node E of Fig. 1, for example, one observes that of the three potential links between E's neighbours, namely between B, G and H, only one is realised, namely the link between G and H. The clustering coefficient of node E is therefore deemed to be 1/3. The clustering coefficient for vertex E is a local concept—it pertains to E and its locality only. If we average local clustering coefficients over the entire system we obtain the *clustering coefficient* for the entire network, denoted by C. In social networks one's acquaintances tend to know each other, so clustering coefficients tend to be high compared, for example, to a random network. In a client–server network, on the other hand, where each computer is connected to a central sever, there is no clustering as the computers are not directly connected to each other.

There is an alternative measure of clustering, sometimes known as the *transitivity*, commonly used in the sociology literature (Wassermann and Faust 1994). This is a global as opposed to local quantity. Denote by N_Δ the total number of triangles in the network and N_t the number of connected triplets (i.e. paths of length 2), then $C_T = 3N_\Delta/Nt$. The transitivity of a network can be estimated from the degree distribution, we refer to this configuration-model estimate as C_n. It turns out that, while this estimate works reasonably well for some non-social networks, the clustering into communities (see below) means that it typically fails for social networks, for which $C_T > C_n$ (Newman and Park 2003).

A common feature of complex networks is that most vertices can reach other vertices in a number of steps which is small relative to the size of the network (Estrada 2011). This is known as the *small-world* effect (Milgram 1967) and an example of it is the notion of six degrees of freedom discussed earlier. A network is small world if it meets the following two criteria:

- The average path length is similar to the average path length of a random graph of the same size and average degree $\ell \approx \ell_{\text{rand}}$,
- The clustering coefficient is much larger than the clustering coefficient of a random graph of the same size and average degree $C \gg C_{\text{rand}}$.

Social networks are usually small world (Amaral et al. 2000). This property is often used in epidemiology, for example, to model disease transmission in society (Kuperman and Abramson 2001).

An edge in a graph can be assigned a positive or negative sign. In a social or character network, these can be used to distinguish between friendly and hostile interactions, with friendly edges denoted as positive, for example, and hostile edges negative. In the overall network, closed triads with just one hostile edge tend to be disfavoured in real social networks as a single hostile link prompts the opposite node in a triangle to take sides. The propensity to disfavour odd numbers of hostile links in a closed triad is known as *structural balance* (Heider 1946; Cartwright and Harary 1956). This is related to the notion of "the enemy of my enemy is my friend." Structural balance is normally encountered as a dynamical property. For example, it has been observed in the shifting alliances of nations in the lead-up to war (Antal et al. 2006). However, it has also been analysed statically by measuring the abundance of triangles with an odd number of positive edges (see the chapter in this volume titled "Analyses of a Virtual World").

In some social networks, people tend to associate with other people who are similar to themselves (e.g. in terms of ethnicity, religious beliefs, etc.). This tendency is known as *homophily* in sociology. A network in which nodes tend to associate with nodes of a similar degree is termed *assortative* (Newman 2002). Networks that have the opposite property are called *disassortative*. Examples of assortative and disassortative networks are displayed in Fig. 2. It turns out that assortativity is an important property which helps to distinguish social networks from other networks because the former are usually assortatively mixed by degree, whereas non-social networks tend to be disassortative (Newman and Park 2003). In fact, for our analysis, assortativity will play a key role. To investigate the

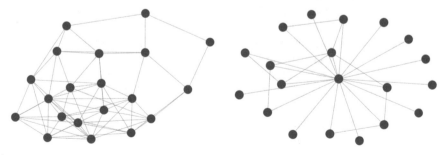

Fig. 2 The network on the *left* illustrates assortativity because nodes of similar degree tend to be linked to each other. This is quantified by a positive correlation coefficient $r_k = 0.6$. The network on the *right* which has, $r_k = -0.6$, has the opposite tendency and is disassortative

assortativity properties of a network, one simply compares the degree of a node to the degrees of its neighbours. Technically, this is done by measuring the Pearson correlation coefficient r_k for the degrees of pairs of nodes connected by an edge.

To test the robustness of a network, vertices can be removed and the size of the giant component measured. This is the largest subset of the network which is connected in the sense that each node can be reached from another traversing network edges. If the giant component fragments quickly then the network is not robust (Albert et al. 2000). The process of removing vertices from the network can be performed in a targeted or random manner. Targeted removal of vertices involves eliminating nodes with the highest degrees, for example. Networks that have power-law degree distributions tend to be robust to random removal of vertices but fragile to their targeted removal (Albert et al. 2000). In the context of the social networks presented in this work, a lack of robustness indicates the network is overly reliant on a few characters. Naïvely, we would expect a social network to be robust due to a lack of disassortativity. For narratives however, this may not be the case as they are often centred on a single protagonist or perhaps a few main characters tie various communities together. Testing the robustness therefore provides information as to whether the tale is hero-centred or society-centred; if the network is held together by only one or two characters connecting with most other nodes, their removal will fragment the network. This will not happen in society-centred stories.

In a network, it is often useful to identify influential or central nodes or edges. One measure of influence is the *betweenness centrality*. This is the total number of *geodesics* (shortest paths) that pass through a given node or edge (Freeman 1977). An edge with a high betweenness centrality has a high probability to be on a shortest path between two other vertices. Therefore it controls the flow of information between various regions of the network. A *clique* is subgraph of three or more vertices wherein each vertex is linked to every other vertex in the clique. Social networks can have large cliques but the larger the clique size, the fewer of them there tends to be. The requirement that each member of a clique be connected to all others is quite extreme. A *community* is a looser sub-collection of connected vertices with dense connections amongst each other and sparser connections beyond that cluster (Newman and Girvan 2004). One possible way to identify communities within a graph is through the Girvan-Newman algorithm (Girvan and Newman 2002). This algorithm removes edges with the highest betweenness as these tend to be maximal when they connect different communities. After each removal, the edge betweenness is recalculated. This process breaks the network down into smaller sub-components as it progresses. Note that community structure is not exclusive to social networks; for example biological, technological and economic networks also exhibit community structure (Fortunato 2010).

Many social networks have already been studied and examples include networks of company directors (Davis et al. 2003), jazz musicians (Gleiser and Danon 2003), movie actors (Amaral et al. 2000), users of online-forums (Kujawski and Abell 2011), and scientific co-authors who have collaborated together on academic papers (Newman 2001). Newman and Park (2003) have demonstrated that social networks are different to other types of complex networks. Thus the general properties of real social networks are well established and well documented and the following properties are typical indicators of them:

- Structural balance
- Community structure
- Small worldness ($\ell \approx \ell_{\text{rand}}$, $C \gg C_{\text{rand}}$)
- High clustering coefficient ($C_T > C_n$)
- Right-skewed degree distribution
- Non-negative assortativity ($r_k \gtrsim 0$).

Non-social networks may exhibit some of these properties but the combination of all seems indicative of social networks in many empirical studies carried out so far. These are therefore the main properties we wish to investigate as we see to map out the characteristics of mythological networks.

Four European Tales: *Njáls Saga*, the *Iliad*, *Beowulf*, and the *Táin Bó Cuailnge*

In this section, we report on the network structures of four iconic European tales: the Icelandic *Njáls saga*, the Greek *Iliad*, the Anglo-Saxon *Beowulf*, and the Irish *Táin Bó Cuailnge*. We chose these because their statuses in comparative mythology are similar to those of Ising or Potts models in statistical physics; they have been widely studied and represent fertile research areas which continue to be investigated.

Njáls saga is one of the *Íslendinga sögur*, or Sagas of Icelanders. These are texts describing events purported to have occurred in Iceland in the period following its settlement in late ninth to the early eleventh centuries. It is generally believed that the texts were written in the thirteenth and fourteenth centuries by authors of unknown or uncertain identities but they may have oral prehistory (O'Donoghue 2004). The texts focus on family histories and genealogies and reflect struggles and conflicts amongst the early settlers of Iceland and their descendants. The sagas describe many events in clear and plausible detail. *Njáls saga*, in particular, is widely regarded as the greatest piece of prose literature of Iceland in the Middle Ages and more vellum manuscripts containing it have survived compared to any other saga. It also contains the largest saga-society network. The epic deals with blood feuds, recounting how minor slights in the society could escalate into major incidents and bloodshed. The events described are purported to take place between 960 and 1020 AD and, while many archaeologists believe the major occurrences described in the saga to be historically based, there are clear elements of artistic embellishment.

The Iliad is an epic poem attributed to Homer and is dated to the eighth century BC. It is set during the final year of the war between the Trojans and a coalition of besieging Greek forces. It relates a quarrel between Agamemnon, king of Mycenae and leader of the Greeks, and Achilles, their greatest hero. Also much debated throughout the years (Wood 1998), some historians and archaeologists maintained that the *Iliad* is entirely fictional (Finley 1954), while recent evidence suggests that the story may be based on a historical conflict around the twelfth century BC interwoven with elements of fiction (Kraft et al. 2003; Korfmann 2004; Papamarinopoulos et al. 2012).

Beowulf is an Old English heroic epic, set in Scandinavia. A single codex survives which has been estimated to date from between the eighth and eleventh centuries. The story relates the coming of Beowulf[1], a Gaetish hero, to the assistance of Hrothgar, king of the Danes. After slaying two monsters, Beowulf returns to Sweden to become king of the Geats (a tribe inhabiting what is now Götaland in Sweden) and, following another fabulous encounter many years later, he is fatally wounded. Although the poem is embellished by obvious fictional elements, such as monsters and a dragon, archaeological excavations in Denmark and Sweden offer support for the historicity associated with some of the human characters (Anderson 1999). Nonetheless, the character Beowulf himself is mostly believed not to have existed in reality (Klaeber 1950; Chambers 1959).

The *Táin Bó Cuailnge (Cattle Raid of Cooley)* is the most famous epic of Irish mythology and describes the invasion of Ulster (in the north of Ireland) by the armies of Queen Medb of Connacht (in the west) and the defence by Cúchulainn. Related to the *Táin* itself are a number of pre-tales and tangential tales (*remscéla*) which give the backgrounds and exploits of the main characters. The *Táin Bó Cuailnge* has come down to us in three recensions. The first has been reconstructed from partial texts contained in *Lebor na hUidre* (the Book of the Dun Cow, dating from the eleventh or twelfth century and compiled at the monastery at Clonmacnoise) and *Lebor Buide Lecáin* (the Yellow Book of Lecan, a fourteenth century manuscript) and other sources. The second, later recension is found in *Lebor Laignech* (the Book of Leinster, a twelfth century manuscript formerly known as the *Lebor na Nuachongbála* or Book of Nuachongbáil, after a monastic site at Oughaval in County Laois). A third recension comes from fragments of later manuscripts and is incomplete. Two popular English translations of the *Táin* (Kinsella 1969; Carson 2007) are mainly based on the first recension, although they each include some passages from the second. We base our analysis on Kinsella's version which therefore, for the purposes of this analysis, serves as a proxy for what is commonly understood as the *Táin Bó Cuailnge.* We sometimes refer to it as the *Táin,* for short.

Analyses of Mythological Networks

The data for the networks were harvested by carefully reading each of the narratives and entering characters' names into databases, meticulously listing the characters they interact with. We defined links as positive ("friendly") when two characters

---

[1] We distinguish *Beowulf* the narrative from Beowulf the character by writing the former in italics.

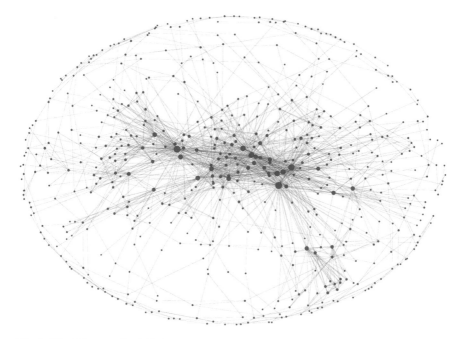

Fig. 3 The full network of *Njáls saga*. Hostile links are in *red* while friendly interactions are represented in *grey*. Heavier edges represent stronger ties between the characters

know each other, are related, speak to one another, or appear in a small congregation together. Links were deemed negative ("hostile") if two characters meet in combat. We also assigned a "weight" to the links between characters, based on how often they encounter each other in the narratives.

The full networks for the four narratives are depicted in Figs. 3 (*Njáls saga*), 4 (*Iliad*), 5 (*Beowulf*), and 6 (*Táin*). In each case, positive edges are depicted in grey and negative ones in red. The properties of the full networks and of the positive subnetworks are listed in Table 1 where N denotes the number of nodes (characters); $<k>$ is the mean degree, γ is the power that best fits a power-law degree distribution, ℓ is the mean path length, C is the clustering coefficient, C_T and C_n are the transitivity and its naïve estimate respectively, G_c is the proportion of sites in the giant component, r_k is the degree assortativity and Δ is the percentage of triangles that have an odd number of negative edges.

We observe is that there is little difference between some of the measured properties of the full networks and their positive subgraphs. In particular, the mean degree $<k>$, the power γ, the mean path length ℓ, the maximum path length $\ell_{\max}$, the clustering C, differ by less than 10 % between the full and positive networks. This may be interpreted as indicating that, even though conflict is an essential element of each narrative, they are stories about human relations, driven primarily by positive interactions. We also observe that each network is small world with $\ell \approx \ell_{\mathrm{rand}}$ and $C \gg C_{\mathrm{rand}}$. (*Njáls saga* is the least small world with ℓ about a third larger than ℓ_{rand}.

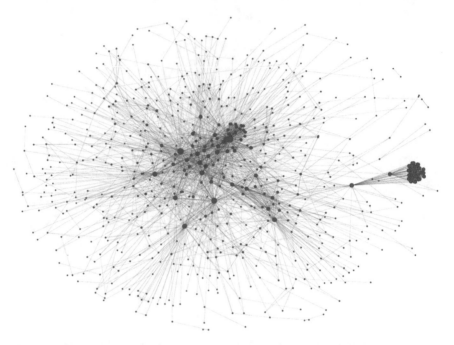

Fig. 4 The full *Iliad* network. *Grey* edges represent positive (friendly) interactions and *red* indicates negative (hostile) ones

The network, however, has an unusually long diameter of $\ell_{max} = 24$ which affects the average path length. This is due to the extensive genealogies present within the text.) The full networks are structurally balanced with a minority of triangles containing an odd number of negative edges. They also have community structures to varying degrees; the society depicted in Njáls saga, for example, appears more homogeneous to those of the other tales.

The commonality of these features across the various networks suggests they are good candidates to be universal properties of mythological networks. We are also interested in distinguishing features and we discuss these for each narrative individually. The mean degree $<k>$ is strongly dependent on the size of the network (in fact it is $2L/N$ where L is the number of edges), so it is not a good comparability measure. Instead we focus on the size and robustness of the giant component, the community structure and the assortativity.

Njáls saga is the largest of the sagas of the Icelanders with $N = 575$ nodes in total. It contains many more negative edges than the other Icelandic sagas with 224 of the 1612 edges purely hostile. The full network is shown in Fig. 3 with the hostile edges represented as red lines and the friendly interactions in grey. It has a high clustering coefficient, given by $C = 0.40$, and its transitivity is almost three times its naïve prediction. *Njáls saga* is robust, removing 5 % of the most connected vertices keeps the giant component at 77 % of its original size. The overall network has

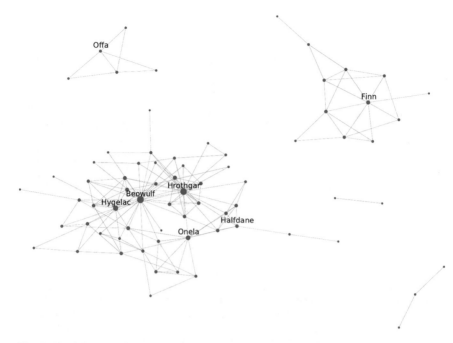

Fig. 5 The full *Beowulf* network. *Grey* edges represent positive (friendly) interactions and *red* indicates negative (hostile) ones

Fig. 6 The full network for the *Táin*. *Grey* edges represent positive (friendly) interactions and *red* indicates negative (hostile) ones

Table 1 Network statistics

Network	N	$<k>$	γ	ℓ	ℓ_{rand}	ℓ_{max}	C	C_{rand}	C_T	C_n	G_c	r_k	Δ
Njál	575	5.6	1.6	5.1	3.7	24	0.40	0.01	0.26	0.09	100	0.01	10
Njal$^+$	564	4.9	1.7	5.4	4.1	24	0.39	0.01	0.28	0.08	96	0.07	–
Iliad	694	7.7	1.7	3.5	3.4	11	0.43	0.01	0.45	0.13	99	−0.08	3
Iliad$^+$	640	7.3	1.7	3.8	3.5	10	0.44	0.01	0.58	0.11	86	0.09	–
Beowulf	72	4.6	2.4	2.4	2.9	6	0.57	0.06	0.37	0.17	69	−0.12	13
Beowulf$^+$	68	4.1	2.1	2.5	2.5	6	0.56	0.06	0.40	0.17	68	−0.07	–
Táin	422	6.0	2.2	2.8	3.6	8	0.73	0.01	0.10	0.62	99	−0.33	12
Táin$^+$	405	5.5	2.1	3.0	3.0	8	0.74	0.01	0.10	0.53	93	−0.32	–
Beowulf* (positive)	67	3.5	2.2	2.8	2.9	7	0.52	0.05	0.42	0.11	66	0.01	–
Táin* (positive)	405	2.9	2.1	4.0	5.6	8	0.36	0.01	0.35	0.03	47	0.03	–

Here, *Beowulf* means the full *Beowulf* network; *Beowulf$^+$* is the positive subnetwork; and *Beowulf** refers to the specially modified subnetwork explained in the text. Similar notation applies to the other narratives. N denotes the number of nodes; $<k>$ is the mean degree; γ is the parameter that best fits a truncated power-law degree distribution in the case of *Njals* saga and the *Iliad* networks or a power-law degree distribution for *Beowulf* and the *Táin*; ℓ is the mean path length; C is the clustering coefficient; C_T and C_n are the transitivity and its naïve estimate; G_c is the size of the giant component as a percentage of the number N; r_k is the degree assortativity; Δ is the proportion (in percent) of triangles with an odd number of negative edges

$r_k = 0.01(2)$ but the positive subnetwork is more assortative with $r_k = 0.07(3)$. (The numbers in parentheses here and throughout are error estimates obtained by the jackknife method.) We will shortly see that this value is similar to that of the *Iliad*.

Apart from an unusually homogeneous community structure, these numbers indicate that *Njáls saga* has features of a real social network. O'Donoghue (2004) gives an extensive discussion of the historical reliability of the various Icelandic sagas. They describe many events in clear and plausible detail but it could be that the sagas are fiction framed in such a way as to appear realistic or historical to the modern reader. However, even if the events they portray are fictional, they may play out against a backdrop which includes real history. In other words, it is possible that the structure of medieval Icelandic society may have been preserved in saga form while the events may be fictional. Either way, it is "almost impossibly difficult" to unambiguously distinguish fact from fiction in such sagas (O'Donoghue 2004).

The largest of the four networks considered here is that of the *Iliad* which has $N = 694$ nodes. The full network is depicted in Fig. 4. The *Iliad* network is small world and has higher transitivity than naïvely expected. The giant component of the full network contains 99 % of its nodes. Usually for collaboration networks, the proportion of vertices belonging to the giant component is under 90 % (Newman 2001). Indeed, the corresponding figure for the positive *Iliad* subnetwork is 86 %. This may hint that the positive subnetwork is more appropriate than the full network for exploring the social structure underlying the narrative. The *Iliad*'s positive subnetwork is relatively robust to the removal of nodes with the giant component

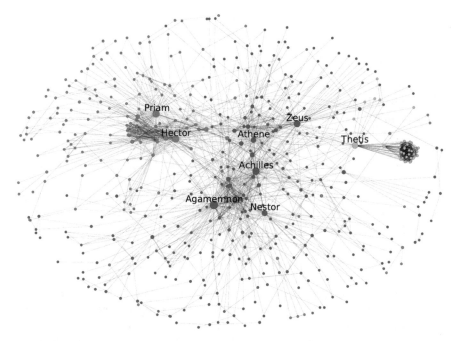

Fig. 7 The Girvan-Newman algorithm detects three communities in the positive (friendly) sub-network of the *Iliad*. Here, the *blue vertices* are from the Greek faction, the *red vertices* are Trojan characters and the *green nodes* are Neirids

being 72 % of its original size when 5 % of the most connected nodes are removed. (For the full network the corresponding figure is 66 %.) The Girvan-Newman algorithm detects three clear communities in the positive subnetwork of the *Iliad* and these are clearly visible in Fig. 7. There, the red vertices are the Greeks, the blue ones are the Trojans and the green nodes are a clique of 34 Nereids (sea-nymphs) who appear to comfort Achilles' mother Thetis. Besides the Greeks, Trojans and Nereids, the gods form the fourth most significant community. (The remaining eight communities have 22 or fewer vertices.) The full *Iliad* network is mildly disassortative with $r_k = -0.08(2)$.

Disassortativity is a feature which is usually characteristic of non-social networks. Indeed, Gleiser (2007) suggested that an analysis of degree correlations reveals the artificial nature of a set of fictional narratives (the so-called "Marvel Universe"). Does their negative assortativity indicate that the interactions depicted in the *Iliad*, *Beowulf* and the *Táin* are artificial? We have seen previously that the positive subnetwork may be more appropriate than the entire network for the study of societal structure. It turns out that the disassortativity observed in some tales is, at least in part, a reflection of the conflictual nature of the stories; many characters are introduced which are killed off virtually immediately by one of the heroes. Such encounters generate links between low-degree victims and high-degree heroes and add to the disassortativity (or reduce the assortativity) of the networks. This again suggests that assortativity as an indicator of the structure of the underlying society may apply to the friendly networks only. In the case of the *Iliad*, the negative subnetwork is strongly disassortative with $r_k = -0.45(5)$ but the positive subnetwork is assortative with $r_k = 0.09(2)$. This is similar to the value for *Njáls saga* and means that the positive subnetwork of the *Iliad* has all the features of a real social network.

The similarity between the network structures of *Njáls saga* and the *Iliad* is bolstered by consideration of their degree distributions. They are both right skewed, well described by truncated power laws and fits deliver similar exponents, namely $\gamma = 1.63(1)$ for *Njáls saga* and $\gamma = 1.69(3)$ the *Iliad*. Their cumulative degree distributions are presented in Fig. 8 where this similarity is visible.

The giant component of the full *Beowulf* network contains less than 70 % of the vertices. This is relatively low compared to the other narratives listed in Table 1 and is because of two sub-tales in the story dealing with events from the past. The giant component is fragile to the systematic removal of vertices in order of highest degree (only 32 % remains after removing the top 5 % of characters) but it is relatively robust to random removal of nodes. Again the transitivity exceeds the naïve expectation. The community detection algorithm finds five components in the friendly network (Fig. 9). All of these features, together with the universal properties identified above, are found in other (real) social networks. In this context, then, it is interesting to observe that the full *Beowulf* network is disassortative with $r_k = -0.12(6)$. Again we turn to the positive subnetwork. In the case of *Beowulf*, however, the assortativity of the positive network is $-0.07(7)$—closer to the non-negative values expected for real social networks but still negative.

As theoretical physicists, we are quite used to playing with models and with data. For example, even though the world we inhabit is four dimensional, theoretical

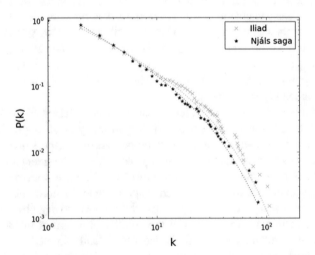

Fig. 8 The degree distribution for *Njals saga* (*black stars*) and the *Iliad* (*green crosses*) and both fitted by truncated power laws represented by the *fitted lines*. The distributions are similar with exponents $\gamma = 1.63(1)$ and $\gamma = 1.69(3)$, respectively

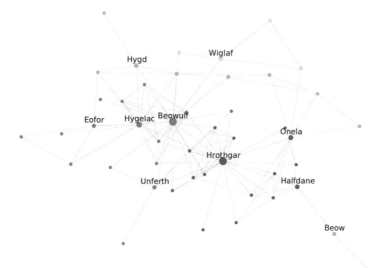

Fig. 9 The positive *Beowulf* giant component. Here the *red vertices* represent Geats and the *green nodes* are the Danes whom Beowulf goes to aid. The *blue vertices* are the Swedes who had been at war with the Geats and the *yellow nodes* are the characters who were involved in the incident with the dragon towards the end of the narrative. The two *grey vertices* are ancestors of Beowulf

physicists are quite happy to change the dimensionality of their models to explore what *would* happen in alternative circumstances. In the present case we recall that archaeological evidence offers some support for the historicity of many of the human characters in *Beowulf* the tale, but Beowulf the character is generally not believed to have been based on reality. We now do what is natural for theoretical physicists but what some humanities scholars and archaeologists might consider radical; we remove the eponymous protagonist from the network. Since Beowulf himself has a high degree and he is connected to many low-degree characters, this has the effect of further reducing the disassortativity in the system. We denote the resulting positive network by *Beowulf**. Indeed, it turns out that *Beowulf** has $r_k = 0.01(9)$, indicating a structure akin to that of a real (or realistic) social network.

The *Táin* is one of the most extensive narratives in Irish mythology. It also has a large proportion of hostile edges.[2] The full network is robust when vertices are randomly removed but the giant component diminishes rapidly when the most connected of vertices are removed by degree. Exceptionally for the networks studied here, the transitivity is significantly *smaller* than the naive estimate. The community

---

[2]The number of characters alone is not very informative in the current context; we are interested in interactions and, for social networks, positive interactions in particular. In the *Táin*, for example, many characters—some named and some unnamed—are introduced to be immediately killed off by Cúchulainn. We keep only a representative amount of named characters. Thus, for the full network, the values of N in Table 1 may be interpreted as minimum sizes.

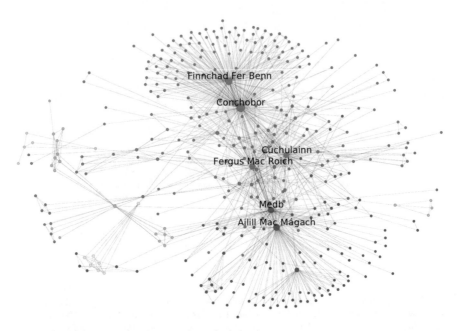

Fig. 10 The Girvan-Newman algorithm applied to the friendly network of the *Táin*. Different coloured vertices represent different communities. The two largest are the Ulster faction (*red*) and the Connacht faction (*blue*)

detection algorithm is applied to the positive network to give a better representation of the factions and the algorithm finds six communities. These communities are shown with different colours in Fig. 10. The two largest communities are the two opposing sides in the conflict. The six most connected characters are named in the figure and the algorithm assigns them all to the correct factions. Fergus Mac Roich however changes factions within the story and spends more of the story in the Connacht (blue in Fig. 10) faction than the Ulster (red) one. The full *Táin* network is strongly disassortative and restricting our attention to the positive network delivers a value $r_k = -0.32(5)$.

In Table 2 we summarise some of the main outcomes of the above analysis. *Njáls saga* has many properties of real social networks. Its positive subnetwork is similar to the overall network. This is because hostile links in the narrative tend to be formed through blood feuds as opposed to open war. In the *Iliad*, *Beowulf*, and the *Táin Bó Cuailnge*, hostile links are instead formed by otherwise unacquainted characters meeting in battle. Although the full network of the *Iliad* is disassortative, its positive subnetwork is assortative. In this sense, the societies depicted in *Njáls saga* and the *Iliad* are quite similar (from a networks point of view). In contrast, both the full *Beowulf* and *Táin* networks and their positive subnetworks are disassortative. In the metric space of comparative mythology, these are also close to each other and remote from *Njáls saga* and the *Iliad*. This is the first major conclusion of the work

Table 2 Summary of some of the distinguishing properties of the positive and adjusted networks associated with the four epics analysed herein

Network	High clustering	Robust	Scale free	Assortative
Njál+	Yes	Yes	No	Yes
Iliad+	Yes	Yes	No	Yes
Beowulf+	Yes	No	Yes	No
Táin+	No	No	Yes	No
Beowulf*	Yes	No	Yes	Borderline
Táin*	Yes	No	Yes	Borderline

Here the + symbols indicate positive subnetworks and the symbols * indicate that the networks have been adjusted as described in the text. High degrees of clustering are signalled by $C_T > C_n$. Whether or not the network is robust to targeted removal of nodes tells us if the tale is society-centred or hero-centred. The positive *Beowulf* and *Táin* degree distributions and their adjusted counterparts are well described by pure power laws and hence are scale free, but *Njals saga* and the *Iliad* are better described by truncated power-laws. In addition to the different properties listed here, the networks are small worlds, have varying degrees of community structure and the full versions are structurally balanced

presented here. MacCarron (2014) gives more extensive comparisons amongst a greater number of epic narratives.

It is tempting to explore these features further and the quantitative approach (coupled with physicists' universal licence to tweak parameters) again allows for speculation not easily accessible through traditional techniques. As we have done for *Beowulf*, one may then ask: what would it take to diminish the disassortativity of the *Táin* also? In particular, does one have to alter the nodal structure of the entire network or, like in the case of *Beowulf*, would some local changes suffice? Figure 10 holds a clue: there we overlap the degree distribution for the *Táin* network with that of *Beowulf*. We observe a remarkable similarity between the two.

This structural similarity between the social networks underlying the narratives is confirmed quantitatively by similar exponents $\gamma = 2.4 \pm 0.2$ and $\gamma = 2.2 \pm 0.1$, respectively. The degree exponents for the positive subnetworks are very similar to these values. The similarity, however, breaks down for the top six most connected characters in the *Táin* character set. They are offset relative to the solid *Beowulf* line. This means that the degrees of these characters are too "large" relative to the characters of *Beowulf*, and, indeed, relative to the other characters in the *Táin*. This hints where the disassortativity of the *Táin*, is located.

To match the *Táin* line with an extended version of that of *Beowulf*, one would have to reduce the degrees of the six anomalous characters in the Irish narrative. This means changing the storyline so that the number of direct interactions these have with other characters is reduced. To illustrate an example of how this might be done, we firstly define a *weak link* as one that occurs when two characters meet only once in the entire narrative. We then speculate that such weak links might be proxy interactions in the tale. For example, the single encounter between Queen Medb and one of her warriors chosen to fight Cúchulainn may rather be interpreted as an encounter between Medb's proxy and that individual. Removing the weak

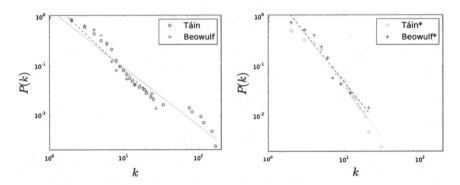

Fig. 11 *Left*—The full degree distributions of *Beowulf* and the *Táin* strongly overlap except for the six characters with the highest degree for the *Táin*. *Right*—the adjusted *Táin** degree distribution overlaps even more strongly with that of the modified *Beowulf**

links associated with the top six *Táin* characters reduces their respective degrees. (It is important to note that what we have done here is to remove weak links—not to remove the characters themselves.) The right panel of Fig. 11 gives the degree distribution for the adjusted *Táin* network, which we refer to as *Táin**. Overlapping it with the degree distribution of the *Beowulf** network shows that the top six points are no longer offset and nearly all the datapoints fall on the same line.

We next examined the assortativity of the *Táin**. The removal of the weakest links of the six anomalous characters has the effect of rendering the entire network (marginally) assortative.

The network remains small world and with $C_T = 0.33$ the transitivity is over eight times higher than $C_n = 0.04$. Thus, removing the weakest interactions of only six characters has the effect of rendering the network more realistic like those of *Njáls saga* and the *Iliad*.

In addition to Medb, Cúchulainn and Fergus Mac Roich, mentioned above, the six anomalous *Táin* characters identified by the above procedure include Ailill mac Mágach (husband of Medb), Conchobar mac Nessa (king of Ulster) and Finnchad Fer Benn (Conchobar's son). In discussing a possible historical basis for the *Táin*, the linguist Kenneth Jackson (1964) demonstrated that the Ulster Cycle of Irish mythology, which includes the *Táin Bó Cuailnge*, preserves an ancient oral tradition which reflects Celtic Irish society. He famously demonstrated that such narratives corroborate Greek and Roman accounts of the Celts and offer us a "window on the Iron Age". In particular, he believed that "the characters Conchobar and Cúchulainn, Ailill and Medb and the rest, and the events of the Cattle Raid of Cooley, are themselves entirely legendary and purely un-historical. But this does not mean that the traditional background, the setting, in which the Ulster cycle was built up is bogus" (Jackson 1964). It is interesting to observe that four of the characters named by Jackson are amongst the six identified as anomalous in the above networks-based approach.[3]

Conclusions

We have presented complex-networks analyses of four famous European mythological epics. By quantifying aspects of the societal structures which underlay these tales we are able to determine elements of similarity between them. For example, the networks are, to various degrees, small worlds and structurally balanced. Overall network properties are dominated by their positive subnetworks, indicating that, the stories are primarily driven by positive interactions between characters. Other network properties can be used to differentiate between the structures of the societies and, indeed, to group them into similarity classes. In particular, the *Iliad* shares many features with *Njáls saga*, while *Beowulf* and the *Táin Bó Cuailnge* are also similar to each other. This result reflects and quantifies the observation that *Beowulf* and the *Táin* are character-centred stories, focussing in particular on the exploits of protagonists Beowulf and Cúchulainn. *Njáls saga* and the *Iliad*, on the other hand, are more aimed at a societal level.

Of all the various network quantities, the assortativity is the one which perhaps most varies between the networks examined. Most non-social networks tend to be disassortative while real-world social networks tend not to have this property. *Njáls saga* is the only narrative analysed here whose full network is assortative. For the *Iliad*, while the full network is slightly disassortative, the positive subnetwork is assortative. In the case of *Beowulf* and the *Táin*, the full network and the full positive subnetworks are disassortative. Coupling expert opinion in the humanities with a certain freedom enjoyed in theoretical physics, we explore the *Beowulf* network obtained by removing its eponymous protagonist. The resulting network is indeed more realistic when compared to social networks. Similarly, local changes in the *Táin* network achieve the same result. Four of the six manipulated nodes coincide with those identified by Jackson and the network that remains echoes the properties of many real world networks.

Acknowledgements We thank Silvio Dahmen and Joseph Yose for discussions which helped inform the content and presentation of this chapter.

---

[3]The question of historicity is an old one and long debated. For example, while Thomas O'Rahilly (1946) objected that such tales have no historical basis whatsoever, Myles Dillon (1948) believed that sagas give "a picture of pre-Christian Ireland which seems genuine."

References

Albert, R., & Barabási, A.-L. (2002). Statistical mechanics of complex networks. *Reviews of Modern Physics, 74*, 47–97.

Albert, R., Jeong, H., & Barabási, A.-L. (2000). Error and attack tolerance of complex networks. *Nature, 406*, 378–382.

Amaral, L. A. N., Scala, A., Barthélémy, M., & Stanley, H. E. (2000). Classes of small-world networks. *Proceedings of the National Academy of Sciences of the United States of America, 97*, 11149–11152.

Anderson, C. E. (1999). *Formation and resolution of ideological contrast in the early history of Scandinavia*. PhD Thesis, The University of Cambridge, Cambridge.

Antal, T., Krapivsky, P. L., & Redner, S. (2006). Social balance on networks: The dynamics of friendship and enmity. *Physica D, 224*, 130–136.

Buldyrev, S. V., Parshani, R., Paul, G., Stanley, H. E., & Havlin, S. (2010). Catastrophic cascade of failures in interdependent networks. *Nature, 464*, 984–985.

Campbell, J. (1949). *The hero with a thousand faces*. Princeton: Princeton University Press.

Carson, C. (2007). *The Táin: A new translation of the Táin Bó Cúailnge*. London: Penguin.

Cartwright, D., & Harary, F. (1956). Structural balance: A generalization of Heider's theory. *Psychological Review, 63*, 277–293.

Chambers, R. W. (1959). *Beowulf: An introduction*. Cambridge: Cambridge University Press.

Cheswick, W., & Burch, H. (1999). Mapping the internet. *Computer, 32*, 97–98.

Davis, G. F., Yoo, M., & Baker, W. E. (2003). The small world of the American corporate elite, 1982–2001. *Strategic Organization, 1*, 301–326.

Dillon, M. (1948). *Early Irish literature*. Chicago: University of Chicago Press.

Dunne, J. A., Williams, R. J., & Martinez, N. D. (2004). Network structure and robustness of marine food webs. *Marine Ecology Progress Series, 273*, 291–302.

Estrada, E. (2011). *The structure of complex networks: Theory and applications*. Oxford: Oxford University Press.

Finley, M. I. (1954). *The world of Odysseus*. New York: Viking.

Fortunato, S. (2010). Community detection in graphs. *Physics Reports, 486*, 75–174.

Freeman, L. C. (1977). A set of measures of centrality based on betweenness. *Sociometry, 40*, 35–41.

Girvan, M., & Newman, M. E. J. (2002). Community structure in social and biological networks. *Proceedings of the National Academy of Sciences of the United States of America, 99*, 7821–7826.

Gleiser, P. M. (2007). How to become a superhero. *Journal of Statistical Mechanics: Theory and Experiment*, P09020.

Gleiser, P. M., & Danon, L. (2003). Community structure in jazz. *Advances in Complex Systems, 6*, 565–573.

Heider, F. (1946). Attitudes and cognitive organization. *The Journal of Psychology, 21*, 107–112.

Holovatch, Y., & Palchykov, V. (2007). *Mykyta the Fox and networks of language*. arXiv preprint: arXiv:0705.1298.

Jackson, K. H. (1964). *The oldest Irish tradition: A window on the Iron Age*. Cambridge: Cambridge University Press.

Kinsella, T. (1969). *The Táin*. Oxford: Oxford University Press.

Klaeber, F. (1950). *Beowulf and the fight at Finnsburg*. Lexington, MA: D. C. Heath.

Korfmann, M. (2004). Was there a Trojan war? *Archaeology, 57*, 36–38.

Kraft, J. C., Rapp, G., Kayan, I., & Luce, J. V. (2003). Harbor areas at ancient troy: Sedimentology and geomorphology complement Homer's Iliad. *Geology, 31*, 163–166.

Kujawski, B., & Abell, P. (2011). Virtual communities?: The Middle East revolutions at the guardian forum: Comment is free. *European Physics Journal B, 83*, 525–529.

Kuperman, M., & Abramson, G. (2001). Small world effect in an epidemiological model. *Physical Review Letters, 86*, 2909.

MacCarron, P. (2014). *A network theoretic approach to comparative mythology*. PhD Thesis, Coventry University, Coventry.

MacCarron, P., & Kenna, R. (2012). Universal properties of mythological networks. *Europhysics Letters, 99*, 28002.

MacCarron, P., & Kenna, R. (2013). Network analysis of the Íslendinga sögur—the Sagas of Icelanders. *European Physical Journal B, 86*, 407.

Milgram, S. (1967). The small world problem. *Psychology Today, 2*, 60–67.

Moore, C., & Newman, M. E. J. (2000). Epidemics and percolation in small-world networks. *Physical Review E, 61*, 5678.

Moreno, J. L. (1934). *Who shall survive? A new approach to the problem of human interrelations*. Washington DC: Nervous and Mental Disease Publishing Co.

Myers, C. R. (2003). Software systems as complex networks: Structure, function, and evolvability of software collaboration graphs. *Physical Review E, 68*, 046116.

Newman, M. E. J. (2001). The structure of scientific collaboration networks. *Proceedings of the National Academy of Sciences of the United States of America, 98*, 404–409.

Newman, M. E. J. (2002). Assortative mixing in networks. *Physical Review Letters, 89*, 208701.

Newman, M. E. J. (2006). Finding community structure in networks using the eigenvectors of matrices. *Physical Review E, 74*, 036104.

Newman, M. E. J. (2010). *Networks: An introduction*. Oxford: Oxford University Press.

Newman, M. E. J., & Girvan, M. (2004). Finding and evaluating community structure in networks. *Physical Review E, 69*, 026113.

Newman, M. E. J., & Park, J. (2003). Why social networks are different from other types of networks. *Physical Review E, 68*, 036122.

O'Donoghue, H. (2004). *Old Norse-Icelandic literature: A short introduction*. Oxford: Blackwell Publishing.

O'Rahilly, T. F. (1946). *Early Irish history and mythology*. Dublin: Dublin Institute of Advanced Studies.

Paczuski, M., & Hughes, D. (2004). A heavenly example of scale-free networks and self-organized criticality. *Physica A, 342*, 158–163.

Papamarinopoulos, S. P., Preka-Papadema, P., Antonopoulos, P., Mitropetrou, H., Tsironi, A., & Mitropetros, P. (2012). A new astronomical dating of Odysseus' return to Ithaca. *Mediterranean Archaeology and Archaeometry, 12*, 117–128.

Scala, A., Amaral, L. A. N., & Barthélémy, M. (2001). Small-world networks and the conformation space of a short lattice polymer chain. *Europhysics Letters, 55*, 594–600.

Schweitzer, F., Fagiolo, G., Sornette, D., Vega-Redondo, F., & White, D. R. (2009). Economic networks: What do we know and what do we need to know? *Advances in Complex Systems, 12*, 407–422.

von Ferber, C., Berche, B., Holovatch, T., & Holovatch, Y. (2012). A tale of two cities: Vulnerabilities of the London and Paris transit networks. *Journal of Transportation Security, 5*, 199–216.

Wassermann, F., & Faust, K. (1994). *Social network analysis*. Cambridge: Cambridge University Press.

Wood, M. (1998). *In search of the Trojan War*. Berkeley: University of California Press.

Medieval Historical, Hagiographical and Biographical Networks

Robert Gramsch, Máirín MacCarron, Pádraig MacCarron, and Joseph Yose

Abstract In recent years, a new method to study narrative texts was introduced, using network analysis. The approach is original and adventurous; instead of focusing on the literary or narrative bases of the texts, it involves extracting data for a formalised analysis of network structures. These are determined from descriptions of events in the texts and their statistical properties are studied using standard network-analysis tools. In this way comparisons between chronologically and geographically different texts are possible. Furthermore, we can compare these textual networks with real social networks, studied by modern sociologists or, indeed, fictional ones. These studies have clearly shown that social-network analysis forms an effective bridge between very different disciplines. It can connect scientists and humanists in joint research; it can depict old research questions in a new light and connect different phenomena belonging to the worlds of nature and culture. The key to this bridge is the understanding of complex systems and their emergent properties. But we are still in the very beginning of exploring these issues and in developing an adequate methodology as we seek to incorporate a number of tools recently developed. Here we attempt to cross the bridge from the humanities side. To this end, we present the results of two studies of medieval sources. Our focus is on visualisation and interpretation of local network properties, an approach which is complementary to complexity analyses. We show that the method can offer powerful augmentation to traditional approaches to the humanities and we outline ways in which these can be developed for the future.

R. Gramsch
Historisches Institut, Friedrich-Schiller Universität Jena, Jena, Germany

M. MacCarron (✉)
Department of History, University of Sheffield, Sheffield S3 7RA, UK
e-mail: m.maccarron@sheffield.ac.uk

P. MacCarron
Department of Experimental Psychology, University of Oxford, South Parks Road, Oxford OX1 3UD, UK

J. Yose
Applied Mathematics Research Centre, Coventry University, Coventry, UK

© Springer International Publishing Switzerland 2017
R. Kenna et al. (eds.), *Maths Meets Myths: Quantitative Approaches to Ancient Narratives*, Understanding Complex Systems,
DOI 10.1007/978-3-319-39445-9_4

Introduction

This chapter is aimed at demonstrating the outcomes of network studies strongly driven by humanities perspectives. Each section involves a comparison of medieval sources. In contrast to the previous chapter by P. MacCarron and R. Kenna, however, here the focus is rather on small networks and the roles of individuals within them. We partition the chapter into two sections. The first is by Máirín MacCarron, a medieval historian at the University of Sheffield, and focuses on texts from late-seventh- and early-eighth-century Britain. The supporting material, which aids the visualisation, was generated by Pádraig MacCarron at Oxford. The second section was authored by German medievalist Robert Gramsch from the University of Jena, and concerns descriptions of the events surrounding the succession to the German throne in 1002. That work was supported by Joseph Yose, a complexity scientist at Coventry's Applied mathematics Research Centre. Besides the comparisons and contrasts discussed within each section, we believe comparisons between these two contributions as well as between this chapter and the previous one are worthwhile. The latter comparison gives insight into the differences between the research questions and modes of thinking emanating from a humanities background and those with a more science-based orientation.

Scientists, for example, are primarily interested in large texts so that they can bring the full power of their statistical tools to bear on the problem. From experience in the physical world, they are acutely aware that sample size determines the amount of information available and, consequently, the level of confidence we have in our calculated estimates. Their interests are in the macro-world: average properties and trends in entire populations or their textual equivalents. Humanities scholars, on the other hand, are more used to operating with small data sets. Indeed, the networks examined in this chapter contain between 21 and 84 network nodes. Such small sample sizes are not well suited to statistical analysis, and any such statistics are presented under that caveat. By way of comparison, a commonly used social network study is that of Zachary's karate club (Zachary 1977) which, although having only 34 nodes, for a time provided a benchmark for community detection algorithms (Girvan and Newman 2002). Animal social networks also tend to have small samples; for example Kasper and Voelkl (2009) studied 70 networks which range in size from 6 to 35 primates. Transport networks also can have small samples, and Derrible (2012) analysed 28 metro networks, half of which have 20 or less nodes. We can glean some information from such small social networks through visualisation and by focussing on the properties of the central characters. Indeed, we hope this chapter demonstrates that humanities scholars with precisely the right expertise can extract information from network visualisations that would probably not be accessible to more road-brush scientists.

Some elements of modern research on dramatic and epic textual corpora already include comparative network-based modelling and interpretation (e.g., Trilcke 2013; Fischer et al. 2015; Jannidis et al. 2015). Network analyses are also increasing in prevalence in medieval research, such as using historical networks based on the links between clergy or political actors (Padgett and Ansell 1993; Malkin 2011; Gramsch 2013; for reviews, see Lemercier 2012; Jullien 2011). In such research, the texts form important sources of descriptions of political and personal constellations, of events and the people involved. They serve as a sort of "quarry" for the mining of data. It therefore seems necessary to study these texts for their network properties in order to gain understanding of their realism and representational accuracy.

We open the chapter with MacCarron and MacCarron's networks-inspired analysis of four texts from Anglo-Saxon England in the early middle ages. This is followed by the Gramsch-Yose contribution in which two tracts of text, covering the same events, from medieval Germany are compared.

Section 1: Networks in History and Hagiography: The Two *Lives* of St Cuthbert and Early Histories of Wearmouth-Jarrow

In this section, we present the findings of a pilot study which examined four Latin texts written in England in the Anglo-Saxon period of the late-seventh and early-eighth centuries. The texts are from the kingdom of Northumbria and involve people and events from that region. Specifically, they concern the monasteries of Lindisfarne and Wearmouth-Jarrow (religious houses were centres of literary production in northern Europe at that time). The texts that we assessed are[1] (see Table 1):

(i) the *Life of Cuthbert* by an anonymous author from the monastery of Lindisfarne;

(ii) the *Life of Cuthbert* by the Venerable Bede (AD 673–735), which was written in the monastery of Wearmouth-Jarrow;

(iii) the *Life of Ceolfrith*, written by an anonymous author based in Wearmouth-Jarrow;

(iv) Bede's *History of the Abbots of Wearmouth and Jarrow*.

Our primary purpose is to compare the networks of (i) and (ii) to each other as hagiographies concerned with the same central character but by different authors. We also compare (iii) and (iv) which cover the same monastery, in the same time period, but are examples of two different genres. Our analysis addresses three main issues: firstly, whether we can use networks to detect or visualise differences

---

[1]The *Lives* of Cuthbert, texts (i) and (ii), are accessible in Latin and English in the volume, *Two Lives of Saint Cuthbert* (Colgrave 1940). The *Life of Ceolfrith* (iii) and the *History of the Abbots* (iv) are available in Latin and English in *Abbots of Wearmouth and Jarrow* (Wood and Grocock 2013). Bede's *Life of Cuthbert* (ii) and the two texts about Wearmouth-Jarrow, (iii) and (iv), are also translated in the collection *The Age of Bede* (Farmer and Webb 1998). Thanks to Peter Darby, Julia Hillner, Dáibhí Ó Cróinín, and the members of the University of Sheffield's Late Antiquity Reading Group for their comments on an early draft of this section.

Table 1 The four texts analysed herein categorised according to their subject matter, genre, authorship and the monastery in which the author was based

Subject matter	Genre, author and base	
Cuthbert (of Lindisfarne)	(i) Hagiography by anonymous monk at Lindisfarne	(ii) Hagiography by Bede at Wearmouth and Jarrow
Ceolfrith/Wearmouth and Jarrow	(iii) Hagiography by anonymous monk at Wearmouth and Jarrow	(iv) History by Bede at Wearmouth and Jarrow

between hagiographies and history; secondly, our investigation illuminates the position of women in the first two texts; thirdly, we shed light on the roles of authors in such texts.

Before turning to the network analyses, a brief comment on the genres of hagiography and history may assist those unfamiliar with their meanings in a medieval context. Genre or type of text was very important for medieval writers, as each had its own rules and conventions. Of the four texts that we examined, three are hagiographies and the fourth is a history (Table 1). Hagiographies are concerned with the lives of those who are considered saints in the Christian tradition: they most often take biographical form and usually have the word *Life* in their title.[2] Hagiographies vary in style and substance but they are identifiable in their focus on the life of a saint and their intention to edify the reader. Their Christian character is clear from the outset as they occasionally refer to portentous signs at their protagonist's birth or during their childhood which are presented as indicators of future sanctity. Where these texts differ most obviously from conventional biography is in the attention paid to the saint's influence *after* their death, with several chapters usually devoted to post-mortem miracles intended to demonstrate their sanctity. Their primary purpose was to promote and commemorate the main protagonist, either to bolster the reputation of someone already regarded as a saint or to boost an embryonic cult.

The history genre follows different conventions, and such works usually contain the word *History* in the title. The medieval understanding of history was defined by Isidore of Seville (AD 560–636) in his famous encyclopaedia, the *Etymologies* (Barney et al. 2006: 67).[3] He stated that a history is a narration of deeds accomplished (*Etymologies* Book i.41); and histories usually concern peoples, rather than individuals, and cover a longer period of time than the lifespan of one person (Book i.43). Medieval writers worked within these conventions and authors frequently produced works in different genres as is the case today. The Venerable Bede, who will play a prominent role below, is an example.[4]

---

[2]For the origins of hagiography and the development of devotion to saints in Christianity, see: White (1998), Brown (1981), and Thacker and Sharpe (2002).

[3]For discussion of the genre of history (often referred to by its Latin name: *historia*), which has its roots in Classical Antiquity, see the essay collection: Deliyannis (2003).

[4]For the differences between history and hagiography, see Fouracre (1990).

Lives *of Cuthbert: Texts (i) and (ii)*

Cuthbert of Lindisfarne (AD 634–687) was one of the most important early Anglo-Saxon saints and was commemorated shortly after his death by his monastic community. The first account of Cuthbert (i) was written by an anonymous Lindisfarne monk sometime around AD 700 (Thacker 1989: 115). About 20 years later, the community at Lindisfarne asked Bede to produce another *Life of Cuthbert* (ii). Bede is the most famous and widely-read writer from the Anglo-Saxon Age and his best known work today is the *Ecclesiastical History of the English People* (Colgrave and Mynors 1999). That the Lindisfarne community asked a monk from another religious house to write an account of Cuthbert indicates the esteem within which Bede was held by his contemporaries. Bede also had a long-standing personal devotion to Cuthbert, as he had previously written a *Life of Cuthbert* in verse, and later included several chapters about Cuthbert in his *Ecclesiastical History*.[5]

These two *Lives of Cuthbert* are structured very differently: the anonymous *Life* divides its material into four books, while Bede presents his in a single book of 46 chapters. However, both are biographical in character as they chart Cuthbert's life from childhood through his monastic career to his elevation to a bishopric in Northumbria, retirement to a hermitage on Farne Island where he died, and posthumous miracles (Colgrave 1940). Bede, who was writing 20 years later, at a time when Cuthbert's fame was continuing to increase, also included additional stories and more post-mortem miracles (Thacker 1989).

In an approach which is new for these types of texts, we analysed the social networks of both. Our data collection focussed on connections between individuals, but we did not weight the links: therefore the network does not distinguish between strong relationships (such as spousal) and those between individuals who happened to meet only once. This approach has the benefit of producing data that is simple and we deemed it appropriate for a pilot study. In Section 2 of this chapter, an alternative approach is introduced.

Bede's account (ii) contains more characters than the Lindisfarne *Life* (i): 59 compared to 49; and (i) and (ii) have 19 characters in common. There were 12 unnamed or unidentifiable characters in (i) and 18 in (ii). If these were discernible, one may have a higher number of shared characters between the two texts. There are 41 identifiable characters in Bede's *Life* and 37 in the anonymous *Life*. The 19 shared characters include clerics and monks in the Northumbrian church, members of the royal family, two angels, as well as Hildmer and his wife. The angels interact only with Cuthbert in each *Life* and are without identifiable gender. There are 78 edges (connections between characters) in the anonymous *Life* (i) and 86 in Bede (ii). Both writers record interactions between Cuthbert and the population of Northumbria. However, for the most part these people are unnamed and unidentifiable, or, with the exception of Hildmer and his wife, when names

---

[5]For discussion of the circumstances surrounding the composition of Bede's prose *Life of Cuthbert*, see, for example: Goffart (1988: 258–296), Berschin (1989), and Thacker (1989).

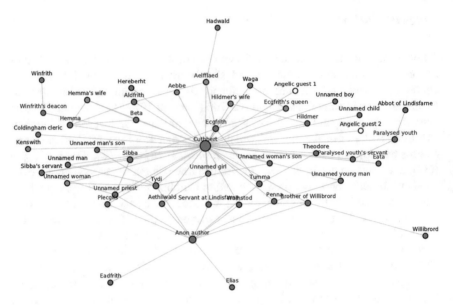

Fig. 1 The social network of the 49 characters in (i) the anonymous *Life of Cuthbert*: *Red nodes* represent female characters, *blue nodes* are male, *white nodes* are angels. Cuthbert is indicated by the *largest node* and the anonymous author is in the *bottom middle*

are provided the sources do not agree. Despite these discrepancies, the graphical configurations appear quite similar, as shown in Figs. 1 and 2. The network properties are also similar, as we shall see.

The Early History of Wearmouth and Jarrow: Texts (iii) and (iv)

The monastery of Wearmouth and Jarrow is familiar to many today primarily because of the Venerable Bede. The first part of this monastery was established at Monkwearmouth in AD 672/3 by Benedict Biscop (AD 628–690). A second house was founded at Jarrow in AD 681/2 at Benedict's instigation, which was ruled by Ceolfrith, and linked to Wearmouth (Wood and Grocock 2013: xxv–xxxii). The establishment of a single monastery in two locations, approximately 9 miles apart, is unusual, but the anonymous writer of the *Life* of Ceolfrith (iii) and Bede in his *History of the Abbots* (iv) both present the community of the two houses living in unity. The texts are concerned with the monastery's early years during the abbacies of Benedict Biscop, Eosterwine, Sicgfrith and Ceolfrith, who ruled both contemporaneously as co-abbots and successively. Both accounts conclude during the abbacy of Hwaetberht, soon after the death of Ceolfrith in AD 716. Much attention has been paid to the relationship between the two accounts (iii) and (iv) and some have speculated that Bede was the author of both texts (McClure 1984),

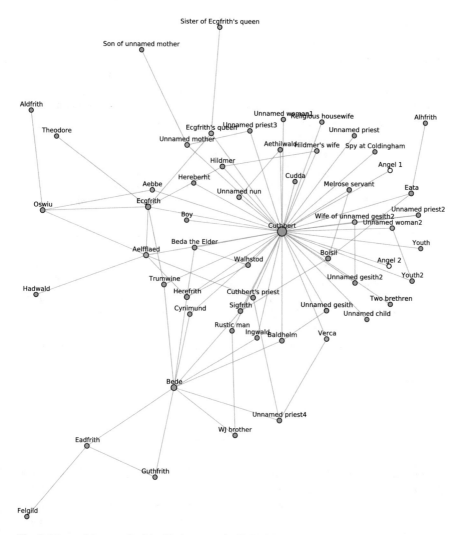

Fig. 2 The social network of the 59 characters in (ii) Bede's *Life of Cuthbert*: *Red nodes* represent female characters, *blue nodes* are male, *white nodes* are angels. Cuthbert is indicated by the *largest node* and Bede, the hagiographer, is the next largest in the *bottom left*

though recent scholarship has rejected this assertion (Wood and Grocock 2013: lxi–xcv).[6]

---

[6]For further information about the relationship between Bede's *History of the Abbots* and the anonymous account of Ceolfrith, and the early history of Wearmouth and Jarrow, see Wood and Grocock's lengthy introduction and consult their extensive bibliography for further material (2013).

The pair of texts (iii) and (iv) is roughly contemporary with the *Lives* of Cuthbert (i) and (ii), but, unlike them, the Wearmouth-Jarrow accounts represent two different genres of writing. Although each has sometimes been regarded as a hagiography (see Wood and Grocock 2013: xxii), Bede's account (iv) is more properly a history. It follows the conventions of historical writing and Bede himself categorised it as such. He provided a list of his extensive writings in an autobiographical note at the end of his magisterial *Ecclesiastical History* (Book v. 24). Organising his output according to genre, he collected his hagiographies under one heading, and grouped his works of history under another. He described his account of the abbots of Wearmouth-Jarrow (iv) as: 'A history of the abbots of the monastery in which it is my joy to serve God, namely Benedict, Ceolfrith, and Hwaetberht, in two books' (Colgrave and Mynors 1999: 571).[7] Bede did not place his *History of the Abbots* with his hagiographies, nor did he call it a *Life* or *Lives* of the abbots.

The anonymous account (iii) has 21 characters while Bede's version (iv) has 23.[8] The texts have 28 and 38 edges, respectively. Of the characters, 11 are shared between the texts: the five abbots of Wearmouth and Jarrow; two kings of Northumbria (Egfrid and Aldfrid); three popes (Agatho, Sergius and Gregory II); and John the Arch-Cantor, whom Benedict Biscop had escorted to England from Rome in AD 678.[9] The difference in character sets is partly because the *Life of Ceolfrith* (iii), in line with hagiographical convention, has much additional information on the protagonist's early years and the end of his life, whereas the *History of the Abbots* tells us more about wider developments in the Anglo-Saxon church and society.

Network Statistics and Genre Identification

A summary of some of the network statistics is given in Table 2. An obvious question is whether any of these can distinguish between hagiographical and historical genres. Notwithstanding the obvious caveat associated with small sample sizes, two measures suggest themselves. The first is the clustering coefficient (transitivity), which, in the history (iv), is larger than in the hagiographies (i)–(iii). This means there are proportionally more closed triads in the history network than in the other three. Indeed, one might expect that two acquaintances of a character are more likely to be connected to each other in a history than in a hagiography. Secondly, but less significantly, the distribution of distances is more varied in (iv) than in

---

[7]Historiam abbatum monasterii huius, in quo supernae pietati deseruire gaudeo, Benedicti, Ceolfridi et Huaetbercti, in libellis duobus (Colgrave and Mynors 1999: 570).

[8]Bede's History of the Abbots has 23 individual characters, one of which represents a group (the unnamed disciples of Pope Gregory I).

[9]For the importance of John the Arch-Cantor, see also Bede, Ecclesiastical History, book iv.18 (Colgrave and Mynors 1999; Hunter-Blair 1990: 171–2).

Table 2 Network statistics for all four texts

	Text (i) by Anonymous	Text (ii) by Bede	Text (iii) by Anonymous	Text (iv) by Bede
Genre	Hagiography	Hagiography	Hagiography	History
Number of characters	49	59	21	23
Mean degree	3.2 (5.6)	2.9 (5.4)	2.7 (3.3)	3.3 (2.9)
Mean path length	2.2 (0.7)	2.4 (0.8)	2.2 (0.7)	2.5 (1.0)
Clustering coefficient	0.1 ± 0.3	0.1 ± 0.2	0.1 ± 0.2	0.3 ± 0.2
Assortativity	-0.4 ± 0.5	-0.4 ± 0.1	-0.6 ± 0.1	-0.4 ± 0.1
Mean betweenness	27.9 (135.0)	37.7 (165.9)	11.5 (34.5)	16.1 (36.8)

The values in parentheses represent the standard deviation from the mean. The error in the clustering coefficient and assortativity are calculated using the Jackknife method

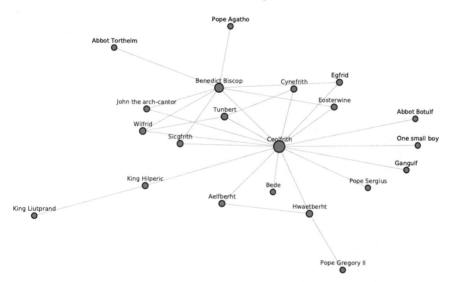

Fig. 3 The network of the anonymous *Life of Ceolfrith* (iii). Only Ceolfrith and Benedict Biscop have more than three links. Ceolfrith is represented by the *largest node*. The *second largest* is Benedict Biscop

any of (i)–(iii). Despite (iv) being a smaller network than (i) and (ii) and having a relatively low mean path length, it has higher standard deviation. These differences are small however and further investigations are required. Indeed, a large-scale quantitative assessment of hagiography and history texts from different parts of Europe is required to test whether bigger data sets can deliver different results.

In the three hagiographies, the titular character is the most connected by a considerable degree (see Figs. 1, 2 and 3). In the anonymous *Life* (iii), for example,

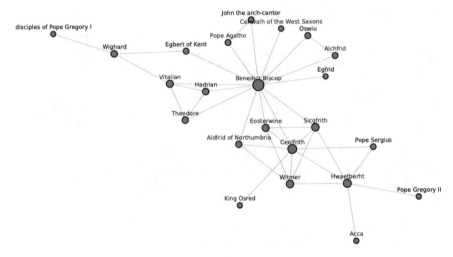

Fig. 4 The network of Bede's *History of the Abbots* (iv). Benedict Biscop is represented by the *largest node* and Ceolfrith is the *second largest*. Eosterwine, Sicgfrith and Hwaetberht are also prominent in terms of their connectedness

Ceolfrith has 16 connections and Benedict has eight and no other character has a degree above three. In Bede's *History of the Abbots* (iv), however, Benedict Biscop is the most connected with 14 links followed by Ceolfrith with eight (Fig. 4). But all five abbots of Wearmouth-Jarrow (Benedict, Ceolfrith, Eosterwine, Sicgfrith and Hwaetberht) are also well connected and located relatively centrally in the network diagram. In fact, and in contrast to the hagiographies, seven characters have a degree above three.

Importance of Gender

In Bede's hagiographical account of Cuthbert (ii) there are 44 men and 12 women. Between these we have 57 male-male interactions; 1 female-female interaction; and 26 male-female interactions. The anonymous *Life* (i), on the other hand, has 39 men and 8 women. There are: 58 male-male interactions; 0 female-female interactions; and 18 male-female interactions. These data are summarised in Table 3.

The conventional view in modern scholarship is that Bede pays less attention to women than the anonymous author. For example, Stephanie Hollis has argued that Bede deliberately curtailed the influence of women in his writings, and her hypothesis has received support from many quarters (Hollis 1992; Blanton 2007: 22–63; and Lees and Overing 2001: 15–39). In examining the *Lives* of Cuthbert, in particular, she suggested that, in contrast to the anonymous author, Bede deliberately suppressed the activity of Abbess Aelfflaed of Whitby (Hollis 1992: 179–207).

Table 3 Numbers (and proportions) of characters broken down by gender and their associated interactions for Texts (i) and (ii)

Text	Characters	Male characters	Female characters	Interactions	Male-male interactions	Female-female interactions	Male-female interactions
(i)	49 (100 %)	39 (80 %)	8 (16 %)	78 (100 %)	58 (74 %)	0 (0 %)	18 (23 %)
(ii)	59 (100 %)	44 (75 %)	12 (20 %)	86 (100 %)	57 (66 %)	1 (1 %)	26 (30 %)

The proportions by gender do not sum to 100 % because a small number of characters have unidentifiable gender

Our observation suggests that the women in Bede's text are more integrated into their social network than women in the anonymous *Life*. As can be seen from Table 3, Bede included a female-female interaction (the connection is between Ecgfrith's queen and her sister), while the Lindisfarne *Life* did not. In addition there is a higher proportion of female characters in Bede as well as a higher proportion of interactions involving females. That the attention paid to females is higher in Bede is further borne out by assessing the betweenness centralities for women in both *Lives* of Cuthbert. In a network, the betweenness centrality of a node gives a measure of the number of shortest paths that the node lies on in proportion to all shortest paths. For information to travel optimally through the network, it is more likely to pass through nodes with a higher betweenness centrality measure. A betweenness of 0 means that node is not on any shortest paths between any other pair of nodes. Here we see that half of the women in Bede's *Life* (ii) have a betweenness centrality greater than zero, while only a quarter of the women in the anonymous *Life* (i) have non-zero betweenness centralities (see Table 4).

By way of contrast, in (i) only 44 % of male characters (17 nodes) have a non-zero betweenness, and only three of these have higher betweenness than Aelfflaed. Similarly, in Bede (ii), 41 % of male characters (18 nodes) have non-zero betweenness, with Aelfflaed again having the fourth highest centrality. The most central character in each network is Cuthbert whose betweenness centrality is 953 in (i) and 1278 in (ii). (Women do not feature in either of texts (iii) or (iv), so we do not have analogues of Tables 3 and 4 for these works.)

Our research suggests that women collectively play a greater part in the social network of Bede's account (ii) than in the anonymous *Life* (i). As noted Aelfflaed is the fourth most connected person in both texts, which is unsurprising from the perspective of Anglo-Saxon society, as she was sister to two successive kings of Northumbria and abbess of the influential monastery of Whitby.[10] However, Aelfflaed's prominence, especially in Bede's text (ii), has, to our knowledge,

---

[10]Whitby was a famous double-monastery, for both monks and nuns, and traditionally ruled by an abbess; it is thought to have been the location of the famous Synod of Whitby, held in AD 664, to determine the method of calculating Easter in the Northumbrian Church; it was a burial place for several members of the Northumbrian aristocracy; and a centre of education and literary production: see Johnson (1993), Thacker (1998), and Lapidge (1999: 472–473).

Table 4 Betweenness centralities for women in Texts (i) and (ii)

Anonymous *Life of Cuthbert* (i)			Bede's *Life of Cuthbert* (ii)		
Name of female character	Degree	Betweenness	Name of female character	Degree	Betweenness
Aelfflaed	4	46.5	Aelfflaed	6	93.0
Ecgfrith's queen	3	0.5	Ecgfrith's queen	3	54.0
Aebbe	1	0	Unnamed mother	3	54.0
Hemma's wife	3	0	Aebbe	2	24.6
Hildmer's wife	2	0	Verca	2	17.7
Kenswith	1	0	Wife of unnamed gesith2	3	0.5
Unnamed girl	2	0	Hildmer's wife	2	0
Unnamed woman	2	0	Religious housewife	1	0
			Sister of Ecgfrith's queen	1	0
			Unnamed nun	2	0
			Unnamed woman1	1	0
			Unnamed woman2	2	0

Contrary to arguments in contemporary scholarship, the data suggests that females in Bede's text (ii) are more integrated into their social network than women in the anonymous *Life* (i)

not been identified in previous scholarship and demonstrates the merits of using networks to aid the analysis of these texts. This also allows us to contribute to an important historiographical debate concerning the treatment of women in these sources, particularly the argument that Bede reduced the role of women in his writings. Our analysis shows that Bede was not suppressing the role of women relative to the writer from Lindisfarne, indeed as Bede included more women and a greater percentage of the women in his *Life of Cuthbert* (ii) have a betweenness centrality above zero, we may speculate that he presents a more realistic picture of Anglo-Saxon society—or, at least, of Anglo-Saxon social networks—than the writer from Lindisfarne. This has wider implications for the study of Bede, both for his presentation of women and for his depiction of society. A full analysis of the social networks of his *Ecclesiastical History* along with a similar assessment of other contemporary Anglo-Saxon hagiographies could therefore be very revealing.

Role of the Author

The role of the author within each text is also an important issue. In Texts (i) and (ii) each author asserts his credibility by citing their sources for various anecdotes that are presented in the works. This serves to insert the hagiographer into the social network of the hagiography. Although this is obvious to any reader of the texts, it is striking that the authors are the most connected people in these networks after the person about whom the text was written. In particular, Bede frequently referred to himself in his *Life of Cuthbert* (ii) to demonstrate the credibility of his witness to Cuthbert's sanctity. Bede did not, however, insert himself into the social network of his *History of the Abbots* (iv). This is perhaps a little surprising as he lived in the monastery for much of the period concerned and was part of the community's social network. He may not have seen the need to assert himself in the same way when writing for his own community in what is a work of history.

In contrast, the authorship of (iii) is an interesting question and we do not know if the anonymous author inserted himself into the *Life of Ceolfrith* (iii). As mentioned above, it has been suggested that Bede wrote this text (McClure 1984). Although recent scholarship overwhelmingly rejects that suggestion, (see, e.g., Wood and Grocock 2013), it is interesting to reflect on what evidence our network approach provides. In our analysis of texts (i) and (ii) we have shown that Bede paid more attention to women than the anonymous hagiographer in (i). In (iii), however, women do not appear despite it covering Ceolfrith's early life when clearly women must have been present. Given the strength of our first observation, we might expect that, were Bede the author, women should appear in (iii) too. It could be argued that their absence indicates that Bede is not the author of (iii). Further evidence for this comes from the observation that Bede is the second most connected character in the hagiography (ii). But he is not the second or even third most connected in the hagiography (iii). [These considerations do not apply to (iv) because that is a history and not a hagiography.] Thus, the network-visualisation analysis delivers the same conclusion as traditional techniques (Wood and Grocock 2013).

Indeed it remains possible that the anonymous author does feature in the text. We suggest that he may be the otherwise unidentified 'small boy', who, we are told, was a student of Ceolfrith's and a priest in the monastery at the time of writing, and was eager to promote the life of his former abbot (*Life of Ceolfrith*, chapter 14). An alternative theory is that this boy is Bede, indeed he has recently been described as 'confidently identifiable as Bede' (Holsinger 2007: 162). However, the suggestion that he could be the anonymous author, who was clearly devoted to Ceolfrith's memory, cannot be ruled out. (If the small boy were identified as Bede it would reduce the number of nodes in Fig. 3.)

Section 2: Networks in High Medieval German Historiography: A Comparison Between Thietmar of Merseburg and Adalbold of Utrecht

Introduction

The purpose of this section is to introduce and evaluate methods which can be used for network analysis of historiographical texts. To demonstrate them, a clear and simple example is used. It is not the objective of this study to gain new insights into the historical events described or into the texts in general. Rather it is to show how the network analytical approach basically works and which cognitive possibilities it would offer within the frame of larger studies.

Concretely, we focus on two German historiographical works from the early eleventh century: the *Chronicle* of Bishop Thietmar of Merseburg and the *Vita Heinrici II Imperatoris* of Utrecht's bishop Adalbold (Ottonian Germany, ed. Warner 2001; Adalbold von Utrecht, ed. Schütz 1999). We consider only excerpts from these works which refer to a specific chain of events, namely the struggle for the German throne in the year 1002.

Our first aim is to compare the networks of political actors according to the narrations in the two chronicles. We will examine to what extent, viewed through the lens of network analysis, the descriptions of Thietmar and Adalbold match. We have to take into account that Adalbold had Thietmar's account to hand when drafting his *Vita Heinrici*. Therefore, one might expect that Adalbold's account is at least as dense as that of Thietmar, and possibly even more accurate since he had additional information. Can network analysis confirm this? And to what extent are differences in the narrative strategies of the authors reproduced in the network models?

A second aim of our study is methodological. In gathering the information from Thietmar's chronicle, we employed two different data-harvesting techniques; we used different rules for the determination of the protagonists (nodes) and their relations (links). One approach is the same as in MacCarron's and Kenna's original papers—disregarding any knowledge external to the texts and, as with Section 1, taking them on their own merits. These networks are *unweighted*, to use the standard parlance (MacCarron and Kenna 2013). The other approach is informed— or weighted—by expert knowledge of medieval Germany from the outset. We term this *hermeneutical*. It is important to note that the distinction between the two approaches refers to data harvesting only; once the data are gathered they can be statistically processed using either approach and interpreted in the light of, and in parallel to, expert humanities-based knowledge. Our second aim is therefore to investigate the extent to which different rules for the determination of the protagonists and their inter-connections affect the resulting network models. What types of data-harvesting should be applied in future studies?

Background to the Texts and the Events Described

We should start with a brief description of the German throne crisis of 1002. The crisis was triggered by the unexpected death of Emperor Otto III in January 1002 near Rome, aged only 21. His body was brought to Aachen where he was buried on 5 April. As Otto died childless, the problem of his successor arose and several princes proclaimed their ambitions to the throne. The first of them was *Duke Henry of Bavaria*, the future *King Henry II*. He was a great-grandson of the first Ottonian ruler, Henry I, and thus the next of kin of the deceased. In February 1002 Duke Henry received the funeral procession of Otto III in Polling near Augsburg and sought support for his candidacy. However, according to Thietmar's account, he didn't meet with the approval of the princes led by Archbishop Heribert of Cologne. These princes declared themselves, by majority, in favour of the Swabian *Duke Hermann II*. Simultaneously in Saxony, the *Margrave of Meissen, Ekkehard I*, staked his claims to the throne. He had been a close friend of Otto III. Although he enjoyed the support of some, such as the Saxon Duke Bernhard, powerful enemies in the Saxon nobility opposed him. These prevented his election and ensured that the Saxons agreed to Henry II as the new king. Ekkehard set out to the west of the kingdom in order to establish contact with Hermann II. However, on 30 April 1002 he was murdered by Count Henry and Udo from Katlenburg in Pöhlde in an unrelated case of private revenge.

Duke Henry of Bavaria then took the initiative. He rushed with an army of loyal followers to Mainz where he was elected king by some of the German nobility on 6 June 1002. His most important ally, Archbishop Willigis of Mainz, crowned him immediately. Thus, a *fait accompli* was established which Duke Hermann II couldn't undo, although he still had strong and loyal followers, especially in the west of the Reich. A direct military confrontation between Heinrich and Hermann failed to manifest itself, although there were proxy engagements in southwest Germany which tended to result in favour of Hermann's supporters. Nonetheless, the newly crowned King Henry II steadily gained the recognition of all leading princes by making a royal tour ("Krönungsumritt") through Thuringia, Saxony and Lower Lorraine. On 8 September 1002 he was crowned at the "right" place, Aachen, by the "right" Archbishop, Heribert of Cologne. On 1 October Duke Hermann finally conceded. This was essentially the end of the crown struggle.

Thietmar's description of these dramatic events leaves nothing to be desired in terms of detail and credibility. It is one of the classic reports of a German "Königserhebung" (the process of appointing a king) in the Middle Ages. Adalbold's account of Henry's "Königserhebung" is also relatively detailed. However, in comparison to Thietmar, his treatment is more liberal and there are a number of omissions which have a clear tendency to favour Henry II. For instance, Adalbold softens Thietmar's report on the events in Polling to create the impression that the funeral procession of Otto III was more favourable to Henry II from the beginning. The episode of Ekkehard's throne candidacy is almost completely ignored—just a brief note of Ekkehard's assassination in Pöhlde remains. Hermann's struggle for the throne is severely curtailed as well.

Even without using network analytical methods, the quite different character of each report is evident to the reader. While Henry II is the main focus and the hero for both authors, Thietmar painted a picture of an initially uncertain struggle in which Duke Henry asserted himself with luck, skilful manoeuvring and successful political networking. In contrast, Adalbold described Henry's victory as an *ab initio* safe bet, with Henry's enemies relegated to auxiliary players or nearly completely written out. Although he had basically the same information available, he obviously was interested in creating a panegyric instead of a historically accurate account. This corresponds to the typical characteristics of the literary genre of Adalbold's work, the royal biography (as opposed to a history).

We considered network-analytical modelling of both texts as potentially suitable for accurately visualising this general finding. Due to the fact that this is a novel and unusual approach to medieval studies, the process should be described below in more detail.

The Methodologies of Gathering Data

The first step of data harvesting consists in identifying the actors or players (nodes) and the ties (edges) between them. On the surface that seems to be very easy and non-ambiguous, but in practise the process raises questions.

The simplest procedure is at the level of "scene structure" of a text. The Romanian mathematician Solomon Marcus already described this very elementary method in the 1970s as follows:

> Let us imagine a spectator who is during the theatrical performance just able to observe the actors' entries and exits and recognize each character, so to distinguish two different kinds of characters from each other. The amount of information, received by every viewer, we should call the scene structure of the play (Marcus 1973, p. 289).

Such a data set cannot, of course, represent the content of a play, novel or historical work, which is obviously much more complex. But it is a controlled process allowing mathematical comparison with other narrative texts. Already this very simple procedure involves a certain act of interpretation, at least in regard to the distinct identification of the actors. But the network-analytical researcher aims to model more information contained within the texts. This opens further scope for interpretation and we have to acknowledge that there are different ways to render a narrative text for an adequate network model. One may, for example, define positive (friendly) and negative (hostile) links between actors in the following way (see chapter "A Networks Approach to Mythological Epics").

Friendly Links are formed between two characters if they speak directly to one another; know each another; are in the same immediate family (siblings or parents); or are present together in a small congregation.

Hostile Links are made between characters if they physically fight or are at war with one another. If two characters argue or are aggressive to one another, this is not sufficient to constitute a hostile edge.

The placing of a single link between characters following the above rather mechanistic rules may be called an *unweighted* approach. It is already considerably refined in comparison to Solomon Marcus' approach in that the discrimination between positive and negative links makes the model more realistic.

There are various ways to weight the network. For example, one could weight links in proportion to the number of encounters between pairs of characters in the text. However these procedures, and the mechanistic classification of positivity and negativity, remain in some aspects simplistic. As the historiographical reports of the throne struggle show, hostile behaviour may not be simply the exercise of violence. Already one may interpret the verbal opposition of a throne candidacy as an act of hostility to the pretender. We refer to methods of data gathering which not only reproduce the formal appearance of communicative acts (like knowing each other, meeting each other, fighting each other using violence), but also reflect the intentions behind them (political support, impairment), as the *hermeneutical* approach. Such a procedure is more interpretative and admittedly cannot be described in a few sentences.

To demonstrate the differences between the unweighted and the hermeneutical approaches, we discuss one example from Thietmar's chronicle in detail, namely the description of an assembly of princes in Frohse (book IV/52 = Ottonian Germany, p. 188s):

> Meanwhile, having learned of their lord's premature death, the leading men of the Saxons sadly convened at the royal estate of Frohse, held as a benefice from the emperor by Count Gunzelin. Giselher, archbishop of Magdeburg, along with his suffragan bishops, Duke Bernhard, the margraves Liuthar, Ekkehard, and Gero, and the great men of the region pondered the condition of the realm. When Margrave Liuthar realized that Ekkehard wanted to exalt himself over them, he called the archbishop and the worthier part of the magnates outside for a secret discussion. During this discussion, he proposed that they swear an oath to refrain from electing a lord and king, either as a group or individually, until a meeting could be held at Werla. This was agreed and consented to by all, with the exception of Ekkehard. Angry at being held back somewhat from the royal dignity, he shouted: 'Oh Count Liuthar, what do you have against me?' Liuthar responded, saying: 'Indeed, have you not noticed that your cart lacks its fourth wheel?' Thus the election was interrupted, and the saying of the ancients was confirmed: to delay for one night is to postpone for a year, and that means to defer for a lifetime.

Which personal interrelations can we deduce from this short report? The following two sociograms (Fig. 5) show, how the unweighted and the hermeneutical approaches lead to very different network models. The small boxes (nodes) represent the actors who are mentioned in the text, the links between them represent the friendly and hostile edges, which can be deduced from the narration.

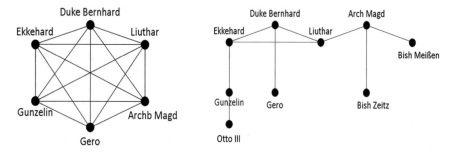

Fig. 5 The meeting at Frohse in 1002 according to Thietmar IV/52. The *unweighted* and *hermeneutical* methods deliver the left and right sociograms, respectively (the *red line* marks a conflictual or hostile link)

The unweighted approach of MacCarron and Kenna pictures this passage simply as a positively associated cluster of six named princes (left panel of Fig. 5). A negative link between Ekkehard and Liuthar is not set because Liuthar's intrigue is not a hostile act in the strict sense of violence or open threat. If we retranslate this unweighted data set back into a text, the narration of Thietmar may essentially be reduced to only one sentence: "The princes … meet each other in a congregation."

In contrast, the hermeneutical approach seeks to model content oriented as precisely as possible with the text (right panel of Fig. 5). In addition to the persons mentioned by name, the bishops of Zeitz and Meissen may be introduced. Their participation at the meeting can be deduced from the text because they are Giselher's of Magdeburg suffragan bishops who, according to Thietmar, were accompanying their metropolitan bishop. Between them and Giselher positive relations are set in the context of a hierarchical subordination due to which they follow him as an act of allegiance (in network analytical terms, a subordination can best be modelled by a star structure). In an analogous manner the secular magnates are assigned to their chief, the Duke Bernhard of Saxony. These attributions are without doubt acts of interpretation, which are however oriented with the wording of the source which is by no means arbitrary in its listing of names. A special position is occupied by Graf Gunzelin who hosted the meeting. He was Ekkehard's brother so that it can be assumed that he wanted to promote primarily Ekkehard's candidacy. Therefore, a positive link is set between him and Ekkehard. Another positive link is directed to the deceased Emperor Otto III, from whom Gunzelin had the royal court of Frohse as a benefice. However, the real key position in the narrative is held by Markgraf Liuthar. As Thietmar explicitly describes, he establishes at the meeting a connection to Archbishop Giseler, in order to prevent Ekkehard's candidacy. Thus Liuthar is on the one hand a broker between the secular and the ecclesiastical greats; on the other hand he is Ekkehard's opponent which is represented by a hostile link. Since its action is politically decisive—in the subsequent narrative his role as a preventer of Ekkehard's candidacy is further specified—it seems justified in regard to the content that we speak here of an open hostility.

The network model derived using the hermeneutical approach requires specialist knowledge in the field of medieval studies as well as knowledge of the modelling options offered by the network approach. While one may claim it follows the text in a manner that the flat unweighted approach cannot, it is not an incontestable objective process. But one should not be quick to reject it because of this; its structure depends on the questions being asked. Different sociograms derived using different interpretations may each be meaningful in its own way.

The aggregation of all the small sub-networks which are displayed in the various chapters of the Thietmar chronicle leads to an overall network. This network constitutes a model of the entire personal constellation of the struggle for the German throne of 1002 as described by Thietmar. As expected, due to the different methods of data gathering, the models have quite different appearances and the unweighted and hermeneutical networks are shown in Figs. 6 and 7, respectively.

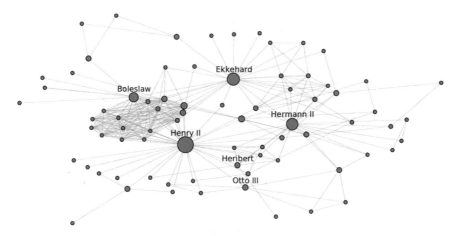

Fig. 6 The unweighted network of the German princes in 1002 according to the Chronicle of Thietmar of Merseburg

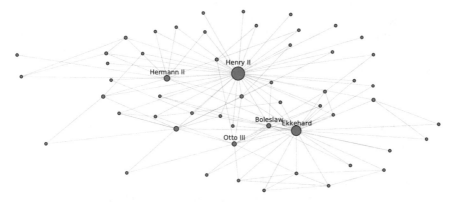

Fig. 7 The hermeneutical network of the German princes in 1002 according to the Chronicle of Thietmar

In the sociograms, the actors are located in a manner which shows important nodes and clustering as clearly as possible. The size of each node is determined by the normalised betweenness centrality (denoted by C) of the actors and, for simplicity, the edges do not distinguish between positive and negative links. The two sociograms differ even in their size. The unweighted one has 84 nodes and 284 edges while the hermeneutical one consists of 55 nodes and 122 edges. Even the ranking of the players varies: Henry II ($C = 0.605$) and Ekkehard ($C = 0.317$) clearly dominate in the hermeneutical record (Fig. 7). They have relatively less prominence in the unweighted approach (with $C = 0.379$ and $C = 0.213$, respectively, in Fig. 6). More-over, in the unweighted approach, there appears other (relatively) important actors, namely Hermann II ($C = 0.186$) and Boleslaw Chrobry ($C = 0.111$). Particularly striking is the denser clustering in Fig. 6. A fully connected cluster of 17 actors (including Henry II) is manifest, which has no counterpart in Fig. 7. The structure is nothing more than a reflection of a scene, in which Thietmar enumerates a total of 17 participants in the by-election of Henry II in Merseburg (including Henry II). According to the unweighted approach, this meeting is considered in terms of the existence of positive relations between all participants, generating 136 dyads at one stroke! The hermeneutical approach interprets the message of Thietmar in a more guarded manner and the election of Henry II is modelled as the establishing of positive relations between Henry and his voters present in Merseburg, i.e. just 16 star-shaped dyads outgoing from Heinrich are set.

It should be stressed once more that there is no simple "right" or "wrong" model. If we consider an assembly primarily as a communication network, it makes sense to model it as a fully connected cluster. If we look, on the other hand, at the political decision arrived at (Henry's election), it is more appropriate to represent the assembly as a star in which there is a clear hierarchy with the elected king in the centre and his princely supporters on the periphery. In principle, the researcher is quite free to choose an unweighted or hermeneutical approach; this decision mainly depends on the particular research question. Of course, the researcher must use the selected method consistently, especially to compare material between narrative texts or with "real" social networks. Here we continue our investigations using the hermeneutical approach because it promises a more representative depiction of the political constellations of the year 1002.

Comparison of the Networks of Thietmar and Adalbold Using the Hermeneutical Approach

The hermeneutical network for Adalbold's texts is depicted in Fig. 8 and is to be compared to Thietmar's network of Fig. 7. Thietmar's is not only bigger but it is also more complex than Adalbold's network. Adalbold's report puts Henry II at the centre, with an overwhelmingly high betweenness centrality ($C = 0.926$). Although in Thietmar's chronicle Henry II takes also a dominant position ($C = 0.605$), a second centre is also looming in the account: the person of Ekkehard ($C = 0.317$). The exceptionally high level of interest Thietmar has in the events in Saxony is very clearly reflected in the network analytical result. Duke Hermann II of Swabia, however, the other major opponent of Henry II, appears very marginal both from Thietmar's perspective ($C = 0.092$) and Adalbold's ($C = 0.077$).

These results may not be so surprising for an expert reader; all these findings can equally be made by a close reading. Nevertheless, in line with the main objective of the survey to discuss the network analytical methodology on a clear and simple example, this outcome is markedly instructive. It becomes apparent that historiographical (or more in general: literary) texts can be usefully "measured" mathematically: Which persons acting have central relevance, how dense is the net of the described interactions, how large the complexity of the plot consequently is?

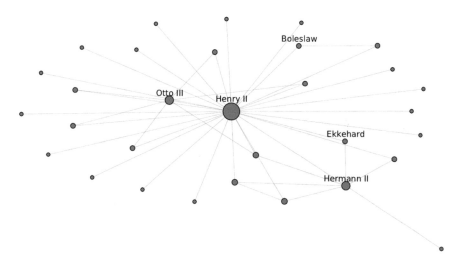

Fig. 8 The hermeneutical networks of the German princes in 1002 according to Adalbold's '*Vita Heinrici*'

In the given case the graphical visualisations convincingly show how differently two authors with a congruent level of information can portray the same events. The distinction between the literary genres—one a chronicle and the other a royal biography—becomes visible at a glance. This is an advance at least in didactical regard. In addition to that we can suppose that the benefit of such a methodology will be much greater if we work with large text corpora. New research questions arise: Does a royal biography also sketch strong secondary characters? Which level of complexity can the historical networks reach? Can we uncover additional insights into the structural properties of the described personal networks, for instance with regard to the existence and the extent of different political factions?

Figure 9 shows the political balance of power at the climax of the succession crisis in the summer of 1002. In this case an algorithm was used (Gramsch 2013) which on the basis of the principle of structural balance divides the personal network into groups which are internally free of conflicts. For the sake of comparability, fixed places have been assigned to the actors in the sociograms. These places nearly match their topographical position. To improve the legibility of the graph, only the most important actors are labelled by names.

The geographical focuses of the reports are now easy to recognise. Adalbold is more accurate in his account of the circumstances in Bavaria (bottom right) where he enumerated the followers of Henry II. However, the conditions in Saxony appear far less important to him which explains the relative emptiness of the upper right quadrant of the right panel of Fig. 9. Thietmar listed here a lot more actors but amongst them were indeed a number of politically insignificant persons. For example, Ekkehard's household knights forming part of a "ring" of actors (in green) on the right side of the left panel of Fig. 9.

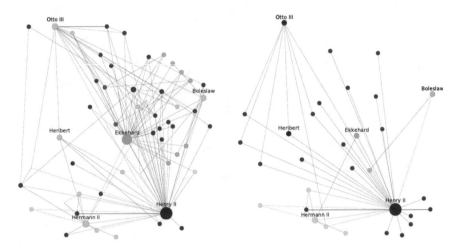

Fig. 9 The hermeneutical networks of the German princes 1002 according to the Chronicle of Thietmar of Merseburg (*left*) and Adalbold's "*Vita Heinrici*" (*right*). Node positions nearly match their geographical locations. Node colours indicate political partisanship; link colours indicate friendly (*blue*), neutral (*dotted blue*) and hostile links (*red*)

The colouring of the nodes corresponds to the division of the network into several groups, determined by the analytical software. They are separated by various conflicts which are represented by red lines. If we consider the distribution of colours in Fig. 9, the big difference between Thietmar's and Adalbold's reports is particularly clear. Adalbold's narrative network is completely dominated by a blue group which is associated with the main actor, Henry II. Only a tiny group of followers (marked turquoise) of the rival candidate, Duke Hermann II, is opposed to the new king's widespread network of supporters. On the other hand, Ekkehard is completely isolated (marked in grey in the right panel). Besides, the dyad of Margrave Henry of Schweinfurt and Duke Boleslaw of Poland stick out as a small group (in green in the right panel); Adalbold here refers already to the protagonists of the uprising of 1003. According to Thietmar, the blue group led by Henry II is significantly smaller. In earlier periods of the conflict (before Ekkehard's assassination and Henry's recognition by the Saxons on the so-called by-election of Merseburg the 24 July 1002) this group is even outnumbered. Ekkehard has his own followers (represented in green in the left panel) although they consist only of low ranking (knightly) actors. The turquoise group of Hermann's (potential) followers is significantly greater than that one described by Adalbold and spreads over all parts of the empire. It is particularly striking that the deceased Emperor Otto III is assigned to this cluster. It points to a circumstance which is repeatedly emphasized by medieval studies: the power shift of 1002 represented a real change in policy, an exchange of the leading elites and a redefinition of policy guidelines, for instance in relation to Poland (Schneidmüller and Weinfurter 1997). This complex scenario is largely hidden in Adalbold's report where even Archbishop Heribert appears as Henry's II follower.

Summary and Conclusions

Here we presented the results of two pilot studies, each carried out from strongly humanities-based perspectives. The first examined the social networks underlying four short texts from early Anglo-Saxon England. Our investigation illuminates the position of women in medieval hagiographical social networks, and allows us to challenge the prevailing view that Bede's *Life of Cuthbert* pays less attention to women than the anonymous *Life*. We also raised questions about the place of authors in hagiographies. The hagiographers in the *Lives of Cuthbert* inserted themselves into the social network of their texts and it is possible that the anonymous author of the *Life of Ceolfrith* did likewise.

The second pilot study compared two historiographical reports on Henry's II coronation in Germany in the year 1002. Although both reports are based on the same information, they exhibit a number of significant structural differences. Adalbold's report is completely focused on the successful candidate, Henry II, who reigns supreme in the centre of a large network of supporters while his competitors appear only marginally. Adalbold's network has a unipolar and politically homogeneous structure. Thietmar, on the other hand, draws a very different picture which leads to doubt over the inevitability of Henry's success. His network appears bipolar and highlights the fact that the crown struggle of 1002 was a quarrel between Henry II and the nobility who were leading under the reign of the deceased Emperor Otto III.

Many of these findings are not new from the point of view of medieval studies. But their reproducibility by means of network analysis confirms the practicability of this new methodological approach. In many respects, it can substantiate existing knowledge, as the acknowledged differences between texts may be reflected in their network properties. In addition, we can observe considerable additional benefits of this methodology: firstly, the network-analytical models can illustrate historical findings very clearly and convincingly and, secondly, they illuminate some hitherto underexposed aspects. They can inspire new questions of familiar material, when used alongside traditional methods of investigation.

Thus, despite the very small sample sizes, our investigations suggest that network-visualisation is a promising approach to illustrate narrative strategies of medieval historians through network analytical methods. Medieval historiography is frequently characterized by an emphasis on personal relationships, and Wolfgang Christian Schneider speaks in this context of a "personal-relational understanding" which formed the historical perception of historians at that time (Schneider 1988, p. 40ss.). The actors do not appear as "closed, individual person[s] but as included in a personal texture of relationships" (p. 30), making these sources an appropriate field for network-theory oriented research.

A second focus of our study was to demonstrate that different methods have different knowledge potentials. A formalistic unweighted approach like that in the first part of this chapter facilitates data gathering and could perhaps be automated in the future. However, when applied to the second data sets, it could be viewed as coarsening the information contained in the sources. It was argued that source-critical interpretive modelling methods (the hermeneutical approach) can map the text more accurately and can achieve improved results in terms of the historical interpretation. However, they have not yet been tested sufficiently and they make high demands on the processor who has to combine historic and network theoretical knowledge. It will be worthwhile in the future to use both methods of data gathering and to tailor them to the aims of the analysis.

Bibliography

Barney, S. A. et al. (Trans.). (2006). *The etymologies of Isidore of Seville*. Cambridge: Cambridge University Press.

Berschin, W. (1989). Opus deliberatum ac perfectum: Why did the Venerable Bede write a second prose Life of St Cuthbert? In G. Bonner, D. Rollason, & C. Stancliffe (Eds.), *St Cuthbert, his cult and his community to AD 1200* (pp. 95–102). Woodbridge: Boydell.

Blanton, V. (2007). *Signs of devotion: The cult of St Æthelthryth in Medieval England, 695–1615*. University Park, PA: Pennsylvania State University Press.

Brown, P. (1981). *The cult of the saints: Its rise and function in Latin Christianity*. London: University of Chicago Press.

Colgrave, B. (Ed. & Trans.). (1940). *Two lives of Saint Cuthbert*. Cambridge: Cambridge University Press.

Colgrave, B., & Mynors, R. A. B. (Eds. & Trans.). (1999). *Bede's ecclesiastical history of the English people*. Oxford: Oxford University Press (1st edn., 1969).

Deliyannis, D. M. (2003). *Historiography in the middle ages*. Leiden and Boston: Brill.

Derrible, S. (2012). Network centrality of metro systems. *PloS One, 7*(7), e40575.

Farmer, D. H., & Webb, J. F. (Trans.). (1998). *The age of Bede* (1st ed., 1965). Harmondsworth: Penguin.

Fischer, F., Kampkaspar, D., & Trilcke, P. (2015). *Digitale Netzwerkanalyse dramatischer Texte*. Accessed 10 July, 2015, from https://www.academia.edu/11531933/Slides_zum_Vortrag_Digitale_Netzwerkanalyse_dramatischer_Texte_._DHd-Tagung_2015_in_Graz_25._Februar_2015

Fouracre, P. (1990). Merovingian historiography and merovingian hagiography. *Past and Present, 127*, 3–38.

Girvan, M., & Newman, M. E. (2002). Community structure in social and biological networks. *Proceedings of the National Academy of Sciences, 99*(12), 7821–7826.

Goffart, W. (1988). *The narrators of barbarian history (AD 550–800): Jordanes, Gregory of Tours, Bede, and Paul the Deacon*. Princeton: Princeton University Press.

Gramsch, R. (2013). *Das Reich als Netzwerk der Fürsten. Politische Strukturen unter dem Doppelkönigtum Friedrichs II. und Heinrichs (VII.) 1225–1235 (Mittelalter-Forschungen, 40)*. Ostfildern: Thorbecke.

Hollis, S. (1992). *Anglo-Saxon women and the church: Sharing a common fate*. Woodbridge: Boydell.

Holsinger, B. (2007). The parable of Caedmon's 'Hymn': Liturgical invention and literary tradition. *The Journal of English and Germanic Philology, 106*, 149–175.

Hunter-Blair, P. (1990). *The world of Bede* (2nd ed.). Cambridge: Cambridge University Press.

Jannidis, F., Krug, M., Puppe, F., Reger, I., Töpfer, M., & Weimer, L. (2015). *Automatische Erkennung von Figuren in deutschsprachigen Romanen*. Accessed 10 July, 2015, from http://gams.uni-graz.at/o:dhd2015.v.011

Johnson, M. (1993). The Saxon monastery at Whitby: Past, present, future. In M. Carver (Ed.), *In search of cult: Archaeological investigations in honour of Philip Rahtz* (pp. 85–89). Woodbridge: Boydell and Brewer.

Jullien, E. (2011). Netzwerkanalyse in der Mediävistik. *Probleme und Perspektiven Wirtschaftsgeschichte, 100*, 135–153.

Kasper, C., & Voelkl, B. (2009). A social network analysis of primate groups. *Primates, 50*(4), 343–356.

Lapidge, M. (1999). *The Blackwell encyclopaedia of Anglo-Saxon England*. Oxford: Blackwell.

Lees, C. A., & Overing, G. (2001). *Double agents: Women and clerical culture in Anglo-Saxon England*. Philadelphia, PA: University of Pennsylvania Press.

Lemercier, C. (2012). Formale Methoden der Netzwerkanalyse in den Geschichtswissenschaften: Warum und Wie? *Österreichische Zeitschrift für Geschichtswissenschaften, 23*, 16–41.

MacCarron, P., & Kenna, R. (2013). Network analysis of the íslendinga sögur—The Sagas of Icelanders. *European Physical Journal B, 86*, 407.

Malkin, I. (2011). *A small greek world: Networks in the ancient mediterranean (Greeks overseas)*. Oxford and New York: Oxford University Press.

Marcus, S. (1973). *Mathematische poetik*. Frankfurt a.M.: Athenäum-Verlag.

McClure, J. (1984). Bede and the life of ceolfrid. *Peritia, 3*, 71–84.

Padgett, J. F., & Ansell, C. K. (1993). Robust action and the rise of the Medici. *American Journal of Sociology, 98*, 1259–1319.

Schimank, U. (2000). *Handeln und Strukturen: Einführung in die akteurtheoretische Soziologie (Grundlagentexte Soziologie)*. Weinheim and München: Juventa-Verlag.

Schneider, W. C. (1988). *Ruhm, Heilsgeschehen, Dialektik: Drei kognitive Ordnungen in Geschichtsschreibung und Buchmalerei der Ottonenzeit (Historische Texte und Studien, 9)*. Hildesheim: Olms.

Schneidmüller, B., & Weinfurter, S. (Eds.). (1997). *Otto III.—Heinrich II.: eine Wende? (Mittelalter-Forschungen, 1)*. Sigmaringen: Thorbecke.

Schütz, M. (1999). Adalbold von Utrecht, Vita Heinrici II imperatoris: Übersetzung und Einleitung. *Bericht des Historischen Vereins für die Pflege der Geschichte des ehemaligen Fürstbistums Bamberg, 135*, 135–198.

Thacker, A. (1989). Lindisfarne and the origins of the cult of St Cuthbert. In G. Bonner, D. Rollason, & C. Stancliffe (Eds.), *St Cuthbert, his cult and his community to AD 1200* (pp. 103–22). Woodbridge: Boydell.

Thacker, A. (1998). Memorialising gregory the great: The origin and transmission of a papal cult in the seventh and early eighth centuries. *Early Medieval Europe, 7*, 59–84.

Thacker, A., & Sharpe, R. (Eds.). (2002). *Local saints and local churches in the early medieval west*. Oxford: Oxford University Press.

Trilcke, P. (2013). Social network analysis (SNA) als Methode einer textempirischen Literaturwissenschaft. In P. Ajouri, K. Mellmann, & C. Rauen (Eds.), *Empirie in der Literaturwissenschaft* (pp. 201–247). Münster: Mentis-Verlag.

Warner, D. A. (Trans.). (2001). *Ottonian Germany: The chronicon of Thietmar of Merseburg*. Manchester: Manchester University Press.

White, C. (Trans.). (1998). *Early Christian lives*. Harmondsworth: Penguin.

Wood, I., & Grocock, C. (Eds. & Trans.). (2013). *Abbots of Wearmouth and Jarrow*. Oxford: Oxford University Press.

Zachary, W. W. (1977). An information flow model for conflict and fission in small groups. *Journal of Anthropological Research, 33*(4), 452–473.

Peopling of the New World from Data on Distributions of Folklore Motifs

Yuri E. Berezkin

Abstract Over the past two decades, a catalogue of folklore and mythology has been created which now contains approximately 50,000 abstracts of texts from all over the world, with information on the distributions of more than 2000 motifs from almost 1000 traditions. In this chapter, we describe this databank and use it to analyse the distribution of motifs across the New World as well as distributions of selected motifs worldwide. Our results support the hypothesis of a coastal route from Beringia to the territories beyond the ice sheets at the early stage of the peopling of America ca. 15–17,000 years ago. Not only do stark differences exist between sets of motifs recorded in North America, on the one hand, and in Central and South America, on the other hand, but these sets find parallels in different regions of the Old World. The Central-South American set has analogies in the Indo-Pacific belt of Asia, especially in Melanesia. Motifs in North America rather often find parallels in continental Siberia. Because all groups of migrants, from the earliest to the Inuit Eskimo, penetrated North America, but few crossed the deserts of Northern Mexico and the Central American bottleneck, the North American folklore traditions accumulated a greater variety of stories and images and are amongst the richest in the world. South American traditions, in contrast, are more homogeneous. Besides this North–South American dichotomy, other regularities in the geographic distribution of folklore motifs in the New World are revealed. In particular, they are related to the presence of different sets of motifs across the South American southwest and east; to the east and to the west of the Rockies; across the North American northwest and across the main part of the continent. These patterns of distribution of motifs can be correlated with archaeological data.

The peopling of America is one of the greatest events or, better said, processes in prehistory with enormous consequences for subsequent cultural development.

Y.E. Berezkin (✉)
American Department, Peter the Great Museum of Anthropology and Ethnography (Kunstkamera), Saint Petersburg, Russia

Faculty of Anthropology, European University, Saint Petersburg, Russia
e-mail: berezkin1@gmail.com

© Springer International Publishing Switzerland 2017
R. Kenna et al. (eds.), *Maths Meets Myths: Quantitative Approaches to Ancient Narratives*, Understanding Complex Systems,
DOI 10.1007/978-3-319-39445-9_5

71

Reconstructions of this process are mostly based on archaeological and, to a lesser extent, genetic research. Here I suggest another, independent, approach, namely using patterns of the geographical spread of folklore motifs. The resulting research can be best understood by comparing with the data supplied by other historical disciplines.

The Database

I have constructed an electronic database of folklore and mythology that includes information on the distribution of more than 2000 cosmological and etiological motifs, adventure and trickster episodes from almost 1000 traditions worldwide.[1] Each entry in the table is supported by one or more abstracts of text with 50,000 abstracts in total. The format allows the data to be easily checked against textual evidence. Here, "folklore and mythology" mean all kinds of traditional stories and tales, long and short, irrespective of their sacred or profane emic meanings. Ninety-nine percent of the abstracts have been written by the present author after reading texts published in Germanic, Slavic, Romance and Baltic Finish languages. The remainder involves unpublished material or texts, the abstracts for which were prepared by other people. The dataset was originally intended as a tool for research on prehistoric migrations and cultural interactions and initially the abstracts were laconic and described only those parts of texts that contained motifs. Since ca. 2000 they have become ever more detailed and closer to the original texts. As new motifs are identified, some abstracts are rewritten in a more complete form, so the corpus of texts (books, photocopies and pdf-files at my disposal) is periodically reviewed. The textual catalogue with auxiliary files takes the form of an electronic table that records the distribution of motifs (which are arranged in columns) among various traditions (rows).

The main analytical unit is the "motif". This is any episode or image[2] related to, or described in, narratives that occur at least in two (in practice many more) different traditions. Some of the motifs correspond to standard indexes used by folklorists. For image-motifs the standard is Thompson's (1955–1958) index of elementary motifs and for the episode-motifs it is the Aarne-Thompson-Uther (ATU) index of types of international folklore (Aarne 1910; Aarne and Thompson 1961; Uther

---

[1]The textual catalogue is available in Russian at http://www.ruthenia.ru/folklore/berezkin and is upgraded once a year. In 2014 an interactive version containing motifs in English and maps of their geographical distributions was created. However, the automatic transfer of data produced errors some of which are possibly still not identified. Because of this, the site is not yet available to the public but it is hoped that it will be in the near future.

[2]For example, the term 'episode' is perhaps best for the description of narratives but it is not usually well suited to cosmological ideas like 'rainbow as a rope' or 'shadows on the moon as a toad'. Therefore, we introduce the terms 'episode-motif' and 'image-motif' to distinguish different types.

2004). However, both systems cannot consistently be used for our purposes because neither was designed as a tool for historical research and both are Eurocentric.

Thompson's index was originally created with a declared aim to remain independent of any historical issues (Thompson 1932: 2). The desire was to reduce texts to a kind of standard combination of basic units. It is characterised by the fact that, while an expert can easily extract a set of registered motifs from a given text, it is usually impossible to restore the content of a text on the basis of the corresponding set of motifs. Descriptions of the root motifs on which clusters of more particular motifs are based were intentionally deprived of details, wordings like "origin of frog" (A2162), "dwarfs in other world" (F167.2), "self-mutilation" (S160.1) being typical. Particular motifs are, on the contrary, too specific and often created considering one unique text (for example A1730, *creation of animals as punishment*, and A1731, *creation of animals as punishment for beating forbidden drum*). As a result, we have a combination of units, some of which are universal and can be found anywhere and others of which have restricted local distributions. In such a situation the statistical processing of regional sets of motifs is impossible. The application of Thompson's index to South American materials (Wilbert and Simoneau 1992) demonstrated that the system itself can be upgraded if necessary to fit the non-European cultural and environmental peculiarities. However, the worldwide processing of units selected on the basis of Thompson's index would reflect the similarities and dissimilarities between environments and cultures and not between oral traditions themselves.

The tale-type was originally understood as a narrative plot with a more or less precise origin in space and time. This idea was severely criticised (Jason 1970) so now the ATU tale-types mostly play a role of reference points in searches for parallels for particular texts. There are several reasons why the ATU index is inappropriate for historical studies, i.e. for assessing a degree of similarity or dissimilarity between folklore traditions across the world. Being Eurocentric, the power of the ATU index to classify the folklore of sub-Saharan Africa, Siberia, Southeast Asia and Oceania is restricted, while Australia and America are completely beyond its scope (since the spread of ATU tale-types across the Old World took place many millennia after the peopling of the New World and Australia and at the time when contacts between Eurasia and these continents were minimal). Ethnic attribution of texts is systematically provided only for Europe. For other areas it is absent or practically absent, not only in the reference index itself (Uther 2004), but even in some regional indexes that use the ATU system (e.g. El-Shamy 2004; Thompson and Roberts 1960; Ting 1978). There are also problems with the European material itself. In many cases sets of episodes found in particular variants of the same tale-type are so different that to assess the degree of similarity between particular texts of the same tale-type, without consulting the original publications, is impossible. Moreover, mistakes are inevitable if textual sources are only named but neither available nor summarized.

Unlike genes, complex tales are able not only to replicate from one generation to another (accumulating random errors as they do so) but also to exchange constituent motifs between tale-types. If parallels with biology are appropriate, the tale-types development is more like the evolution of Prokaryota than like that of Eukaryote. We refer the reader to the Chapter "Phylogenetics Meets Folklore: Bioinformatic Approaches to the Study of International Folktales" for a discussion.

Were I a folklorist, I would hardly venture to criticize the existing indexes. However, as an archaeologist and initially having in mind the specific task of finding potential parallels between scenes on Moche vases and murals (ca. A.D. 100–800; Berezkin 1981), I began to systematize South American folklore data in my own way. Only later, since the mid-1990s, did I become more intimately engaged in problems of method and theory, being influenced more by Boas and his students (Boas 1895: 329–363, 2002: 635–674; Kroeber 1908; Lowie 1908; Wilbert and Simoneau 1992: 41–45) than by mainstream folklore studies.

Most of the motifs in my catalogue were created by selecting common elements in particular texts and not by comparing these texts with standard types created by others a 100 years ago. Because my purpose was not to suggest a new typology, or to revise existing ones, but to study particular historical problems related to the peopling of the New World, not all potential image-motifs and episode-motifs were included into the catalogue. Motifs which are known universally or only locally were ignored and motifs with a trans-regional or transcontinental distribution were sought. The idea was not only to interpret patterns of geographical distributions of particular units, but primarily to create an extensive database that could be processed statistically. The ultimate purpose of my study is not to reveal functional dependencies between elements of culture (and nature), but rather to detect such processes as prehistoric migrations and cultural contacts and interactions.

The American Folklore Data

We wish to undertake statistical processing of the distribution of folklore and mythological motifs in the New World, to try to answer two principal questions: (i) Which sets of motifs in regions between Alaska and Tierra del Fuego differ most? (ii) Can these differences be related to the cultural inheritances of particular groups of migrants that took part in the peopling of America or are they due to regional factors, which are likely to have emerged later?

To answer these questions, we begin by processing the geographical distribution of motifs in the New World and then compare the American and the Old World traditions. Among 2002 motifs that are currently described in the catalogue, 1316 motifs distributed among 413 ethnic traditions of America and the Chukchi Peninsula (the Chukchi and Asiatic Eskimo) have been selected for the first stage of this work. The catalogue contains other motifs as well but they are not registered in America at all or are registered there only once.

For the Old World, the differentiation between adventure and trickster motifs and cosmological and etiological motifs is important because a large part of the former spread only recently, during the last 2000 years or so (Berezkin 2015: 5). But if we remain inside the New World, such a dichotomy is not crucial because no transcontinental interaction sphere comparable with the "world system" of western and central Eurasia had emerged here before Columbus. Regarding motifs of post-Columbian European and African origin, usually they are easily recognisable by anyone with expertise on the folklore material of both sides of the Atlantic. Accordingly, they are not taken into consideration. Cases of uncertain (native or post-Columbian) origin of the motifs are few (Berezkin 2014, 2016) and practically do not influence the statistical results.

Among folklore traditions selected for the New World, 200 are North American (from the St. Lawrence Yupik to the Seri of Northwest Mexico), 44 are Central American (from the Tarahumara to the Cuna of Panama) and 165 are South American. The Central American traditions share many more common motifs with South American ones than with North American ones, so the combined total of the numbers of motifs in Central plus South America is similar to the total number of motifs in North America. At the same time the number of motifs registered for the best represented North American traditions is higher than for the number of best represented Central and South American motifs. In North America over 150 motifs from our list are registered in 21 traditions (as well as amongst the Chukchi), while in South America they occur in only three such traditions. For the five best represented North American traditions, the numbers of registered motifs are: 276 for Lushootseed; 217 for western groups of Ojibwa; 189 for Menominee; 189 for Five Nations Iroquois; and 186 for Tlingit. For South America the highest numbers are: 168 for Warao; 155 for Paresi; 151 for Napo (with Kanelo); 136 for Kariña of Guiana; and 132 for Barasana (with Taibano and Macuna). These differences are not believed to be related to different levels of research intensity because all of the above mentioned traditions are extensively described in the literature. North American folklore traditions, approximately to the north of 40°N, really are the richest in the world, in the sense that they have the greatest number of easily identifiable and widely distributed motifs outside of the western and central Eurasian interaction sphere with its recent spread of "international folktale", i.e. stories largely deprived of specific features related to particular cultures and easily crossing cultural and linguistic borders.

The Quantitative Approach

We are interested in extracting and quantifying the similarities and differences between various traditions' usages of motifs. To do this, we use the mathematical technique of *Principal Component Analysis*. This is designed to bring to the fore the variations (dissimilarities) in sets of data and to filter out noise and redundant information.

In digitised form, the world folklore and mythology database is a binary table (it consists of zeros and ones) with lines for traditions and columns for motifs. Every tradition is characterised mathematically by long strings of zeroes and ones, representing the absence and presence, respectively, of particular motifs. The extent to which any sets of traditions are similar to each other is a function of the degrees of similarity (correlation) between their representative strings.

The strings of zeros and ones are considered to form vectors in a high-dimensional space of motifs. One considers this space to be spanned by underlying, hidden factors or components. The set of factors forms a simpler, lower dimensional space and the objective is to determine its principal components. There are a number of statistical packages available to do this and a number of articles which introduce the technique (e.g., Shlens 2014).

One firstly centres the data about their means and then computes the covariance matrix. The goal is to find an orthogonal transformation which rotates the basis vectors in such a way as to diagonalise the covariance matrix. Because the off-diagonal elements of this matrix are co-variances, rendering them to zero in this manner is equivalent to minimizing redundancy of information (correlations). The remaining elements, those along the main diagonal, are variances, one for each of the principal components. These have to be rank-ordered to pull out the principal components according to importance. Large values of the diagonal terms hold the large variances in the data and hence most interesting structure or information.

The first direction is called the *first principal component* (PC) accounts for the largest possible amount of variability in the data—and thus contains the largest amount of information about the differences between traditions. The next PC has the next highest variance (and information about differences) but is uncorrelated with (orthogonal to) the first PC. Similarly, each subsequent PC is orthogonal to (i.e., uncorrelated with) the preceding ones and contains the maximum allowed remaining variance. Following this method, each tradition is expressed in terms of a small number of factors, namely the principal components. Usually only two, three or four PC's are required and these only account for a small proportion of the variances—less (sometimes much less) than 20 % of the total. However, this turns out to be sufficient to deliver a convincing differentiation of the traditions from the large numbers of features.

Analysing the American Data

The coefficients for the first PC (which represents 4 % of the variability) are represented in Fig. 1 which contrasts North American traditions with Central and South American ones. Two broad sets of traditions, which we call *complexes*, are identified by the signs of the associated indexes. The border between the two lies across northern Mexico, southern California and the Gulf Coast. The data for traditions in the intermediate zone are incomplete; texts that survived from some of them are so few and fragmentary that less than 15 motifs could be selected. Such traditions (Costanoan, Coastal Miwok, Gabrielino and Paipai among them) showed relatively high negative coefficients and are not represented on the map. The Kiliwa tradition is slightly better known (18 motifs) and is represented as a green square in northern Baja California. Coefficients for those south and central

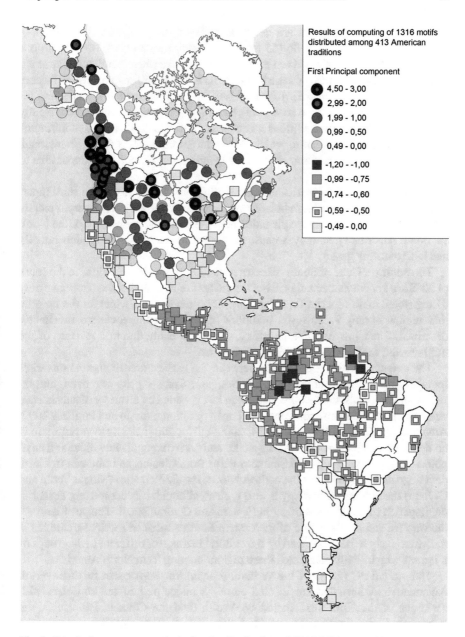

Fig. 1 Principal component analysis for the distribution of 1316 folklore motifs amongst 413 ethnic traditions of the New World. Here, the coefficients associated with the first PC are represented as different shades of *red* (for positive) and *blue* (negative)

Californian traditions which are better represented, like Achomawi (52 motifs), Yokuts (44 motifs) or Luiseño (53 motifs) are very near to zero. Most traditions of the Southwest and of Southern Plains are clearly "northern" and not neutral but they have a greater number of registered motifs too (e.g. 93 for Zuni, 70 for Eastern Keres, 96 for Lipan Apache). It seems that the "slightly southern" tendency for traditions in the intermediate zone between typically North American and typically Central-South American traditions means first of all a lack of typically "northern" motifs. However, the extent to which this is due to incomplete data or how much it reflects the real composition of the corresponding sets of motifs, is impossible to say.

The situation in the American Arctic is different. Traditions of the Inuit (apart from the East Greenland Angmassalik) are excellently known but not very typically North American. A possible explanation is that the Eskimo are outside of the North vs. South America dichotomy because they came to the New World independently and later than any group.

The Southern Cone of South America is another case still. Traditions to the south of 30°S are known inadequately. Even for the best represented of them we have only 51 registered motifs for the Tehuelche and 42 for the Yamana while for the Puelche this number is only 9. However, traditions of the Chacoan people are excellently described. Their low negative indexes are most probably due to a mixture of the northern and southern motifs typical for the area.

The extent to which a difference between the earliest archaeological materials found across the Andean Belt and the Southern Cone, on the one hand, and on Brazilian Highlands, on the other hand, can be explained by a variety of subsistence strategies in different environments is an open question (Araujo and Pugliese 2009; Araujo et al. 2012; Dillehay 2008: 34, 2009; Oliver 2008: 198). However, there is no doubt that a chain of sites with specific and easily recognizable bifacial fishtail points stretches (with a gap in Southern Peru) from Chiapas and Yucatan to Tierra del Fuego and Uruguay. For the earliest known Pacific Belt sites (Monte Verde and El Jobo) the bifacial technology is also typical, although it is completely absent in the Itaparica tradition, the earliest in Eastern and Central Brazil. The peculiarity of the western and southern and of the eastern South American folklore traditions is therefore likely to be influenced by the cultural heritage of different early groups of migrants who initially penetrated these regions coming from North America.

The second PC (2.8 % of the variability or information) contrasts those North American traditions that are located across a major part of the continent with traditions of the Northwest, including Alaska (and the Chukchi Peninsula), the Northwest Coast, the Plateau and the Arctic (Fig. 2). The borderline between the two complexes is quite sharp in the eastern Subarctic (between the Labrador Inuit and the Algonkian traditions) and gradual in northern California and in the western Subarctic. It seems that the revealed tendency results from a combined effect of several historical processes. In the Canadian Arctic and in Greenland the bearers of the northwest set of motifs are the Eskimo. What was a particular role of the recent Thule migration (Sørensen and Gulløv 2012) and of a possible influence of the Paleoeskimo substratum (since 5000–4500 cal. B.P.; Kotlyakov 2014:

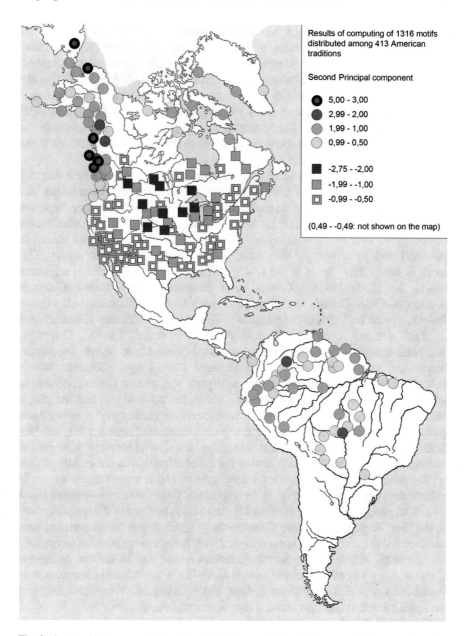

Fig. 2 Geographical representation of coefficients associated with the second PC coming from the analysis of the distribution of folklore motifs amongst ethnic traditions of the Old New World

293–342) is a question apart that does not concern us here. In any case the Eskimo-type cultures had to bring to the eastern Arctic those motifs which initially were spread across western Alaska. Because the Northwest Coast (and Chukchi) traditions are more typically "northwestern" than most of the Eskimo ones (besides the North Alaskan Inupiat which share a lot of specific motifs with traditions of the Chukchi Peninsula, part of them probably spreading recently), the revealed tendency can well reflect Asian cultural links that took place before not only the Thule migration (ca. 800 cal. B.P.) but also before the appearance of Arctic Small Tool tradition (ca. 6000–5000 cal. B.P.).

That the northwestern set of motifs has counterparts in eastern South America is an argument in support of such a suggestion. Traditions with indexes between 0.49 and −0.49 are not shown on the map so the Southern Cone is empty. However, some of the motifs in question are found not only in Amazonia but also among the Fuegians. One of the best examples is the story about a man who turns into powerful bird, or creates it, after which this bird carries away his enemy. This motif is completely absent in the Old World and in North America is registered among the Asiatic Eskimo, Central Yupik (Nunivak Island), Kodiak, Chugach, Aleuts (Unalaska, Umnak, Commander Islands), Bering Strait and North Alaskan Inupiat, Gwich'in, Tagish, Tlingit, Haida, Tsimshian, Carrier and Tillamook and in South America among the Sicuani, Piaroa, Yanomami, Yekuana, Coreguaje, Karijona, Baniwa, Wakuenai, Desana, Witoto, Manao, Mura, Mawe, Parintintin, Muduruku, Tenetehara, Urubu, Moseten, Surui, Cinta Larga, Kamaiura, Waura, Rikbaktsa, Paresi and Yamana. I dare to suggest that such a pattern reflects the spread of motifs that emerged in Beringia before the movement of the first people to the south of the ice sheets. In North America corresponding motifs were largely pushed out by motifs brought by later groups of Siberian migrants but they survived in the northwest, i.e. in the area that during the Last Glacial Maximum was part of Beringia or was peopled in the course of the initial migration by the coastal route.

The third PC (2.6 % of the information) reflects, if it is proper to say so, a twin pattern with the second PC (Fig. 3). In both cases North American—eastern South American parallels seem to correspond to the cultural heritage of the early migrants to the New World but the third PC selects those motifs that in North America have survived not in the northwest but across the North American mainland. The opposite set of motifs, widespread across the American Arctic, Alaska and the Northwest Coast to the north of Vancouver Island, looks like a later Asiatic intrusion. The eastern and central Subarctic are geographically intermediate between the areas occupied by both complexes and demonstrate neutral coefficients.

The fourth PC (2.1 % of the information) is relevant to the situation in North America only (Fig. 4). However, the tendency it reflects is a very important one. The western set of motifs widespread across the Columbia Plateau (with adjacent southern Northwest Coast), California, Great Basin and Great Southwest correlates pretty well with the geographical spread of the Western Stemmed archaeological tradition dated to the post-Clovis time, i.e. to the Younger Dryas (12,800–11,500 cal. B.P.)—Early Holocene (Beck and Jones 2010; Geib and Jolie 2008). Paleoindian Clovis points are found to the west of the Rocky Mountains but predate the Western

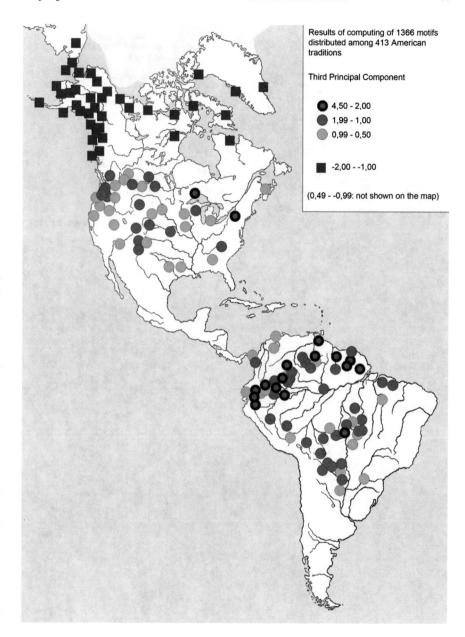

Fig. 3 Geographical representation of coefficients for the third PC

Stemmed archaeological tradition (Fiedel 2014). The western folklore complex is represented by a large series of motifs unknown outside this region or concentrated here much more densely than anywhere else. Thanks to its stemmed points, the Western Stemmed archaeological tradition demonstrates parallels in the Ushki VII complex of Kamchatka dated to 13,000 cal. B.P. (Goebel et al. 2014).

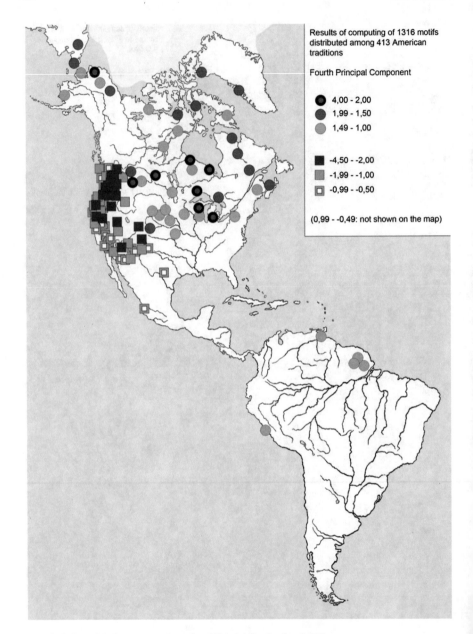

Fig. 4 Geographical representation of coefficients for the fourth PC

The opposite set of motifs reflected by the fourth PC mostly connects traditions of the Algonkian-speaking peoples with the Eskimo and Chukchi traditions of the Bering Strait area. Links between the Algonkians and the Inuits are not direct but are thanks to the Alaskan origins of the Arctic Canadian and Greenland traditions.

The Algonkian—Bering Strait parallels in folklore tales are detailed and exclusive (Berezkin 2010c: 267–269).

It should be noted that there are also two dozen specific parallels shared by the Plateau (mostly Salishan) and Algonkian (mostly Menominee and Ojibwa) groups that can be related to the Plateau origin of the proto-Algonkian speakers (Berezkin 2010c: 265–269). It is a theme apart because on the pan-American scale such parallels are not statistically important. Isolated South American links for the Algonkian-Beringian set of motifs look like chance coincidences.

The American Data from a Global Perspective

The evidence presented above suggests that sets of motifs registered in Central and South America and in North America are the most different from each other. The significance of these differences becomes clear when we analyse the data on a global scale.

Temporally remote transcontinental tendencies in recently recorded folklore material are better seen after eliminating the entropic effect of the western and central Eurasian folktales. After the growth of demographic density and of the intensity of intercultural communication between the Mediterranean Basin, Central, South and to a certain extent East and Southeast Asia ca. 2000 B.P., new forms of folklore, largely deprived of their ethnic specifics and subject to easy borrowing, came to emerge (Berezkin 2015). Though particular stories can be found already in the Ancient Oriental and Greek written sources, most of the genres of the "international folklore" (tales of magic, religious tales, realistic tales) probably emerged later. It is significant that plots of the stories like "Susan and the old men", "Tobias", "Jacob and his brothers" coincide with the plots of Western Eurasian folktales but in the Bible such stories have still a sacred status and can be named "myths".

During their spread across Eurasia, motifs of the "international folktale" over-lapped sets of episodes and images that existed there before. In order to render this earlier "folklore layer" more visible for statistical processing, we removed the motifs most typical for the western and central Eurasian "international folktale". Adventure and trickster episodes with parallels in America and Oceania have been preserved but those which are known exclusively in western and central Eurasia and Africa have not been included. As a result, 1288 motifs out of 2002 remain.

In the plot of Fig. 5, the abscissa (x-axis) represents coefficients of the first PC and the ordinate (y-axis) represents coefficients of the third PC (not the second). Positions of 150 selected traditions from all over the world are plotted (some of the analysed traditions are not represented on the scheme to make it less crowded). Only well known traditions with at least 50 motifs from our list were taken into consideration. For 41 traditions, 150 and more motifs are registered and for 91 traditions, 100 and more. The second PC contrasts the North American and South American traditions, the tendency that was basically demonstrated already on Fig. 1.

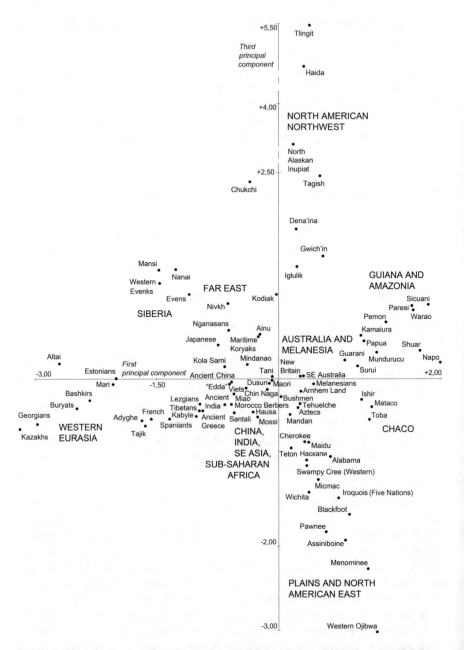

Fig. 5 Position of 150 selected traditions of the world according to distribution of 1288 folklore motifs. The *x*-axis represents the first PC and the *y*-axis represents the third. Here, "Melanesians" mean the Melanesian-speaking groups of the northern coast of Papua-New Guinea; "Papua" means the Trans New Guinea Papuan groups of Papua-New Guinea; and "Tani" means the Sino-Tibetans of eastern and central Arunachal Pradesh

The third PC contrasts traditions of the Plains and North American East with traditions of North American Northwest and demonstrates that in respect to the Siberian traditions the latter stand nearer than the former. For Fig. 5 (150 traditions) the percentage of information corresponding to the first four PC's are 4.3 (First), 3.9 (second), 2.7 (third) and 2.3 (fourth).

The second PC (not on the plot, as mentioned above) combines the South American and the Melanesian traditions into one cluster. It means that all three first principal components cluster the South American traditions together with the Melanesian ones. The first PC demonstrates that on the global scale the South American and Melanesian traditions, on the one hand, and traditions of Western-Central Eurasia (from the Mediterranean Basin to Southern Siberia), on the other hand, are the most different. This difference reflects the opposition between two main sets of cosmological and etiological motifs outside of sub-Saharan Africa. I name them the *Boreal* (or Continental Eurasian) and the *Austral* (or Indo-Pacific) complexes and suggest that they had developed in the Old World after the coming of modern man out of Africa but before the peopling of the New World began (Berezkin 2013). The similarity of Central and South American traditions to the Melanesian ones does not mean, of course, that historically they are related directly. This set of motifs could be widespread in Pacific East Asia during the Last Glacial Maximum and then survived in the parts of the world most isolated from continental Eurasia, i.e. in Melanesia and in South America. Its intrusion into the New World supports the hypothesis of the coastal route by which the first people reached the American mainland 17,000 cal. years B.P. or so starting from the coastal area of Beringia that was later submerged by the marine transgression.

On the scheme (Fig. 5) the Southeast Asian traditions stand near the center of the coordinates because they represent a mixture of the Indo-Pacific and Continental Eurasian motifs. Sub-Saharan Africa is a different case. The mythology of this region is poor, e.g. no etiological motifs related to the origin and appearance of plants were found here at all. Only a dozen etiological motifs, mostly related to the origin of death, had been probably brought from Africa by the early *Sapiences* while the Boreal and Austral complexes of motifs formed later (Berezkin 2007, 2009, 2010a, b).

All the traditions known thanks to the early written sources (the Chinese, Indian, Greek, Scandinavian) are also localised not far from the center of the coordinates. The reason is that the written texts are always deprived of a large number of motifs that we would most probably find if we had access to the corresponding oral traditions. Nevertheless all written traditions of Eurasia stand to the same side from the center of the coordinates as richly represented in our database traditions of the Caucasus, Southern Siberia and Europe.

For Fig. 5 (150 traditions) the proportions of information associated with the first four PC's are 4.3 (for the first PC); 3.9 (second); 2.7 (third); and 2.3 (fourth).

A similarity between the South and Central American and the Melanesian oral traditions is also demonstrated by the spread of 1019 motifs in 41 traditions of the Circum-Pacific region (Fig. 6). Again only traditions that are well represented in the database have been selected. Most of them do not contain "international folklore"

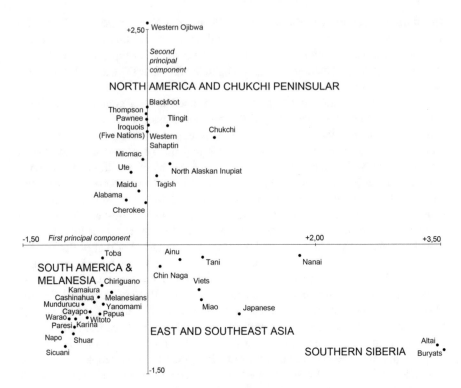

Fig. 6 Position of 41 traditions of the circum-Pacific region according to distribution of 1019 folklore, axis X for first PC, Y for third PC. For "Melanesians", "Papua" and "Tani" see legend for Fig. 5

and those fairytale motifs which are shared by the Altai, Buryats, Nanai, Japanese, Miao and Viets were not processed. The *x*-axis corresponds to the first PC (8.4 % of the information) and the *y*-axis to the second PC (6.8 % of the information). The third and fourth PC's correspond to 5.4 % and 4.1 % of the information, respectively. Not only the resemblance of South American and Melanesian traditions but also their homogeneity deserves attention. In North America the diversity between separate traditions is much higher. The first PC demonstrates that continental traditions of Southern Siberia (the Altai and Buryats) with their typically Boreal sets of motifs stand very far from the American traditions and that this distance is especially big for South and slightly smaller for North America. The Chukchi occupy the place nearby the Alaskan Eskimo and Tlingit, the place of the Nanai is between Southern Siberia and North America.

Conclusions

The data on the geographical distribution of motifs of folklore and mythology both in the New World and on a global level support the hypothesis of the coastal route taken by the earliest migrants to America with subsequent intrusion of groups of continental Siberian origin. Otherwise it is difficult to explain the existence of at least two different sets of motifs in the New World, one of which finds parallels in Melanesia (and in the Indo-Pacific belt of Asia in general) and another in Southern Siberia and Central Asia. This conclusion is supported by statistical processing of the data as well as by patterns of geographical distribution of many dozens of particular motifs.

Besides this major conclusion, there are other regularities in the geographical distribution of folklore motifs in the New World. In particular, they are related to the presence of different sets of motifs across the South American West and South and to the east of the Andes; to the east and to the west of the Rockies; in the North American Northwest and across the main part of the continent. With a reasonable degree of confidence, these patterns in distribution of motifs can be correlated with data from archaeology.

The greater number of motifs registered in the North American traditions in comparison with the Central and South American traditions looks like an effect of a combined heritage of many particular groups of migrants entering the New World, since the initial peopling of Beringia up to the recent movement of the Inuit. By definition, all these groups came to North America but only few (and mostly the earliest ones) crossed the Northern Mexican desert belt and Central American bottleneck.

The existence of such patterns in the geographical distribution of motifs, which can best be explained by the history of the peopling of the New World from archaeological research, opens the possibility to establish *terminus ante quem* for the appearance of particular episodes and images found in recently recorded oral texts. The core of American Indian folklore tradition had to emerge in Asia not later than the Last Glacial Maximum, ca. 20,000 B.P.

Acknowledgment The work is supported by Russian Scientific Fund, grant 14-18-03384.

References

Aarne, A. (1910). *Verzeichnis der Märchentypen*. Helsinki: Suomalainen Tiedeakatemia (FF Communications 3).

Aarne, A., & Thompson, S. (1961). *The types of the folklore: A classification and bibliography*. Helsinki: Suomalainen Tiedeakatemia (FF Communications 184).

Araujo, A. G. M., Neves, W. A., & Kipnis, R. (2012). Lagoa Santa revisited: An overview of the chronology, subsistence, and material culture of Paleoindian sites in Eastern Central Brazil. *Latin American Antiquity, 23*(4), 533–550.

Araujo, A. G., & Pugliese, F. (2009). The use of non-flint raw materials by Paleoindians in Eastern South America: A Brazilian perspective. In F. Sternke, L. Eigeland, & L. J. Costa (Eds.), *Non-flint raw material use in prehistory: Old prejudices and new directions* (pp. 169–175). Oxford: B.A.R. International Series 1939.

Beck, C., & Jones, G. T. (2010). Clovis and Western stemmed: Population migration and the meeting of two technologies in the Intermountain West. *American Antiquity, 75*(1), 81–116.

Berezkin, Y. E. (1981). An identification of anthropomorphic mythological personages in Moche representations. *Ñawpa Pacha, 18*, 1–26.

Berezkin, Y. E. (2007). "Earth-diver" and "emergence from under the earth": Cosmological tales as an evidence in favor of the heterogenic origins of American Indians. *Archaeology, Ethnology and Anthropology of Eurasia, 4*(32), 110–123.

Berezkin, Y. E. (2009). Why are people mortal? World mythology and the "out-of-africa" scenario. In P. N. Peregrine, I. Peiros, & M. Feldman (Eds.), *Ancient human migrations: A multidisciplinary approach* (pp. 242–264). Salt Lake City: The University of Utah Press.

Berezkin, Y. E. (2010a). The emergence of the first people from the underworld: Another cosmogonic myth of a possible African origin. In W. M. J. Binsbergen & E. Venbrux (Eds.), *New perspectives on myth* (pp. 109–125). Haarlem: An African Journal of Philosophy.

Berezkin, Y. E. (2010b). Sky-maiden and world mythology: The dispersal of modern man and the areal patterns of folklore-mythological motifs. *Éditions litteraires et linguistique de l'Universite de Grenoble, 31*, 27–39.

Berezkin, Y. E. (2010c). Selecting separate episodes of the peopling of the new world: Beringian–Subarctic–Eastern North American folklore links. *Anthropological Papers of the University of Alaska, 5*(1–2), 257–276.

Berezkin, Y. E. (2013). *Afrika, migratsii, mifologia. Arealy rasprostranenia fol'klornyh motiviv v istoricheskoi perspektive [Africa, Migrations, Mythology. Areas of Distribution of Folklore Motifs in Historical Perspective]*. Saint Petersburg: Nauka.

Berezkin, Y. E. (2014). Native American and Eurasian elements in post-Columbian Peruvian tales. In K. Antony & D. Weiß (Eds.), *Sources of mythology: Ancient and contemporary myths* (pp. 281–303). Zürich, Berlin: Lit Verlag.

Berezkin, Y. E. (2015). Spread of folklore motifs as a proxy for information exchange: Contact zones and borderlines in Eurasia. *Trames Journal of Humanities and Social Sciences, 19*(1), 3–13.

Berezkin, Y. E. (2016). Children pursued by an ogre: Western and Eastern Eurasian borrowings in the 20th century quechua narratives. *Latin American Indian Literatures Journal, 30*(2), 185–215.

Boas, F. (1895). *Indianische Sagen von der Nordpazifischen Küste Amerikas*. Berlin: Asher.

Boas, F. (2002). *Indian myths and legends from the North Pacific Coast of America*. (D. Bertz, Trans.), Vancouver: Talonbooks.

Dillehay, T. D. (2008). Profiles in Pleistocene history. In H. Silverman & W. H. Isbell (Eds.), *Handbook of South American archaeology* (pp. 29–43). New York: Springer.

Dillehay, T. D. (2009). Probing deeper into first American studies. *Proceedings of the National Academy of Sciences, 106*(4), 971–978.

El-Shamy, H. M. (2004). *Types of the folktale in the Arab world: A demographically oriented tale-type index*. Bloomington: Indiana University Press.

Fiedel, S. J. (2014). Did pre-Clovis people inhabit the parsley caves (and why does it matter)? *Human Biology, 86*(1), 69–74.

Geib, P. R., & Jolie, E. A. (2008). The role of basketry in Early Holocene seed exploitation: Implications of a ca. 9000 year-old basket from Cowboy Cave Utah. *American Antiquity, 73*(1), 83–102.

Goebel, T., Slobodin, S. B., & Waters, M. R. (2014). New dates from Ushki-1, Kamchatka, confirm 13,000 cal BP age for earliest Paleolithic occupation. *Journal of Archaeological Science, 37*, 2640–2649.

Jason, H. (1970). The Russian criticism of the 'Finnish school' in folktale scholarship. *Norweg, 14*, 285–294.

Jon, S. (2014). *A tutorial on principal component analysis*. arXiv preprint arXiv:1404.1100.

Kotlyakov, V. M., Velichko, A. A., & Vasil'ev, S. A. (Eds.). (2014). *Pervonachal'noe zaselenie Arktiki chelovekom v usloviyah menyayuscheisia prirodnoi sredy [Initial human colonization of arctic in changing paleoenvironments]*. Moscow: Geos.

Kroeber, A. L. (1908). Catch-words in American mythology. *Journal of American Folklore, 21*(81–82), 222–227.

Lowie, R. H. (1908). Catch-words for mythological motifs. *Journal of American Folklore, 21*(80), 24–27.

Oliver, J. R. (2008). The archaeology of agriculture in ancient Amazonia. In H. Silverman & W. H. Isbell (Eds.), *Handbook of South American archaeology* (pp. 185–216). New York: Springer.

Sørensen, M., & Gulløv, H. C. (2012). The prehistory of Inuit in Northeast Greenland. *Arctic Anthropology, 49*(1), 88–104.

Thompson, S. (1932). *Motif-index of folk-literature* (Vol. 1). Helsinki: Suomalainen Tiedeakatemia (FF Communications 106).

Thompson, S. (1955–1958). *Motif-index of folk-literature* (Vols. 1–6). Bloomington: Indiana University Press.

Thompson, S., & Roberts, W. E. (1960). *Types of Indic oral tales*. Helsinki: Suomalainen Tiedeakatemia. FF Communications 180).

Ting, N.-T. (1978). *A type index of Chinese folktales in the oral traditions and major works of non-religious classical literature*. Helsinki: Suomalainen Tiedeakatemia (FF Communications 223).

Uther, H.-J. (2004). *The types of international folktales* (Vol. 1–3). Helsinki: Suomalainen Tiedeakatemia (FF Communications 284–286).

Wilbert, J., & Simoneau, K. (1992). *Folk literature of South American Indians general index*. Los Angeles: UCLA Latin American Center Publications.

Phylogenetics Meets Folklore: Bioinformatics Approaches to the Study of International Folktales

Jamshid J. Tehrani and Julien d'Huy

Abstract Traditional narratives, like genes, mutate as they are transmitted from generation to generation. Elements of a myth, legend or folktale may be added, substituted or forgotten, generating new variants that catch on and flourish, or vanish into extinction. Reconstructing these processes has been complicated by the fact that traditional narratives are transmitted via mainly oral means, leaving scant literary evidence to trace their development and diffusion. In this chapter we demonstrate how this problem can be addressed using phylogenetic methods developed by evolutionary biologists. We show how these methods can be used to identify cognate relationships among tales from different societies and eras, reconstruct their ancestral forms, and test hypotheses about how stories evolve. We illustrate how three kinds of phylogenetic analysis can be applied to these problems through two worked examples: Little Red Riding Hood and Polyphemus.

Introduction

Scholars of oral literature have frequently drawn analogies between the characteristics of traditional narratives and biological organisms (Hafstein 2001). Propp (1968) famously proposed that folktales have a consistent structure, or "morphology", which could be dissected into a set of functionally integrated components comparable to bodily organs. Other folklorists (e.g. Gennep 1909; Sydow 1932; Toelken 1969; Thompson 1977), meanwhile, have proposed that traditional narratives have been shaped by the same kinds of evolutionary forces as species: like genes, traditional tales (i.e. myths, legends, folktales, ballads, etc.) mutate as they are

J.J. Tehrani (✉)
Centre for the Coevolution of Biology and Culture, Department of Anthropology, Durham University, Durham, UK
e-mail: jamie.tehrani@durham.ac.uk

J. d'Huy
Institut des Mondes Africains (IMAf), unité mixte de recherche (CNRS—UMR 8171; IRD—UMR 243; EHESS; EPHE; Paris 1 Panthéon-Sorbonne; AMU), Université Paris 1, Panthéon-Sorbonne, France

© Springer International Publishing Switzerland 2017
R. Kenna et al. (eds.), *Maths Meets Myths: Quantitative Approaches to Ancient Narratives*, Understanding Complex Systems,
DOI 10.1007/978-3-319-39445-9_6

transmitted from generation to generation. New elements may be added while others become substituted or forgotten, generating new variants that catch on and flourish, or vanish into extinction. Accordingly, influential writers such as Arnold Gennep (1909) and Thompson (1977) proposed that folklorists ought to classify their material in the same way as zoologists and botanists do, by grouping tales into hierarchically arranged "natural" categories akin to species, genera and families. As Thompson points out "biologists have long since labeled their flora and fauna by a universal system and by using this method have published thousands of inventories of the animal and plant life of all parts of the world. The need for such an arrangement for narratives has been realized for a long time" (Thompson 1977:414).

The most ambitious effort to meet this need has been carried out under the banner of the "historic-geographic method", which was inspired by the work of the Finnish folklorists Julius and Kaarle Krohn in the late nineteenth and early twentieth centuries, and subsequently developed by Antti Aarne, Walter Anderson and Stith Thompson, among others (Thompson 1977). Exponents of this approach developed a taxonomic system for classifying related tales from different cultures known as the "international tale type index" (Uther 2004). An international type is defined as a tale that has "an independent existence" (Thompson 1977:415), which is to say that it comprises a self-contained storyline that can be told on its own (even if it sometimes occurs within a larger cycle of tales, such as the One Thousand and One Nights or The Decameron). International types are recognizable across cultures by virtue of the presence of one or more diagnostic "motifs", which represent "the smallest element in a tale having the power to persist in tradition" (ibid.). Motifs typically comprise specific characters (e.g. a fairy godmother), artefacts (e.g. a magic ring) or episodes (e.g. a man selling his soul to the Devil). To date, over 2000 international tale types have been codified. These are catalogued in the Aarne-Thompson-Uther (ATU) Index (Uther 2004).

Folklorists associated with the historic-geographic school believed that variants of a particular international type could be traced back to an original "archetype" tale that was adapted to suit different cultural norms and preferences, giving rise to locally distinct "ecotypes" (Sydow 1948). They sought to reconstruct this process by assembling all the known variants of the international type and sorting them by region and chronology. Rare or highly localized forms were considered to be of likely recent origin, whereas widespread forms were believed to be probably ancient, particularly when they were consistent with the earliest recorded versions of the tale (Thompson 1977). However, while the historic-geographic method has yielded impressive insights into the evolution of a number of international tale types, it has not led to the fully-fledged science of folklore that it promised. This is for several reasons. First, the applicability of the ATU Index, particularly to non-western traditions is problematic. Since the majority of international types were originally defined in relation to the European corpus, tales from other regions are often difficult to classify because they lack one or more of the key diagnostic motifs, or fall between supposedly distinct international types (Tehrani 2013). A second, related problem with the historic-geographic method is sampling bias. Given that European folklore traditions have been studied far more intensively than any others, reconstructions based on the frequencies and chronologies of variants are likely to be heavily skewed. To overcome these problems, we argue that, as well as looking to evolutionary biology for theoretical models, folklorists have much to gain by importing the kinds of methods that have been used to analyze genetic relationships among species. Specifically, we focus on a set of techniques known as "phylogenetics".

Folklore Phylogenetics

Phylogenetics involves mapping relationships among a group of entities that are related by common descent. In biology, this equates to grouping species into hierarchically nested sets based on increasingly exclusive ties of common ancestry—commonly referred to as "the tree of life". In recent years, phylogenetic approaches have been used to model relationships among various cultural phenomena, including languages (Atkinson et al. 2008; Gray et al. 2010), manuscript traditions (Howe et al. 2001; Roos and Heikkilä 2009) and craft assemblages (e.g. Tehrani and Collard 2002, 2009; Temkin and Eldredge 2007). To reconstruct these lineages, cultural researchers face the same main challenge as their biological counterparts—the problem of missing links. As with the fossil record, there are huge gaps in archaeological and literary knowledge about the past. More often than not, there is no direct evidence about our cultural or biological ancestors. Phylogenetic analysis deals with this problem by using information about the past that has been preserved through the mechanism of inheritance.

The first step in a phylogenetic analysis is to define a set of "characters"—i.e. the basic units of inheritance that can be used to link related taxa. In biology these may be gene sequences or morphological traits. In linguistics they may be lexical items or syntactical features. In material culture they may be specific designs, craft techniques, and so on. In the case of traditional narratives, phylogenetic characters can be derived from the kinds of motifs described above—i.e. characters, objects and episodes that withstand repeated transmission, which folklorists use to identify related versions of the same story (Thompson 1977). Once the motifs have been defined, the state of every character in each variant of a tale is recorded in a matrix. Motifs may be coded as either binary characters that take only two states—e.g. 0 ["absent"] or 1 ["present"]—or as multistate characters that take a variety of expressions—e.g. 0 ["the hero is a boy"], 1 ["the hero is a goat kid"]. 2 ["the hero is a lamb"], etc.

Ideally, the distribution of states for each character would reflect the process of mutation and transmission through which a tale evolved from its original form into all the different variants that exist today. However, it cannot be assumed that all shared motifs have been inherited from a common ancestor. In some cases, similar characters or events may evolve by independent invention, or through borrowing and blending among tales from different narrative lineages. The same issue arises in other fields: distantly related organisms frequently converge on similar morphological solutions to the same kinds of adaptive problems and exchange genetic material (a process that is particularly common in plants and microbes), languages often borrow words from one another (Gray et al. 2010), while both independent evolution and blending play important roles in the evolution of many material culture traditions (Matthews et al. 2011). In all these cases, the key challenge for reconstructing descent histories is to distinguish similarities that are due to inheritance (known as "homologies") from those that are due to other processes (known as "homoplasies").

There are a number of different ways to address this challenge. One of the simplest and most intuitive techniques is known as "cladistics". Cladistic analysis clusters entities (e.g. species, languages or variants of a tale) into hierarchically nested branches known as "clades". A clade represents a group of entities that share evolutionarily novel traits (known as "derived character states") inherited from an exclusive common ancestor. For example, imagine an international tale type X. At some point, a teller of X adds a new ending to the story (e.g. the hero gets married), giving rise to a new version of the story, $X(a)$. This version becomes popular in that teller's community and country, and in time evolves into new versions, with further additions (e.g. the hero and his wife have children). These new versions, let's call them $X(a_i)$ and $X(a_{ii})$, comprise a clade characterized by a particular innovation that is not present in more distantly related versions of X found in other countries $[X(b_i, b_{ii} \ldots), X(c_i, c_{ii} \ldots),$ etc.]. If the evolution of a tale tradition like X conformed exactly to this pattern (i.e. the accumulation of innovations within branching lineages of descent), the task of sorting the variants into clades would be straightforward, since all the similarities among them would be homologous. However, for the reasons explained above, evolution is rarely straightforward. In most cases, we would expect the neat hierarchical pattern of inheritance to be disrupted by independent evolution and/or borrowing, generating *homoplastic* similarities among variants. For example, imagine if $X(a_i)$ included an innovation— for instance, the hero has a magic sword—that is not present in $X(a_{ii})$ but which happens to occur in another variant, $X(c_{ii})$. The distribution of this character conflicts with the distributions of the other two characters mentioned above (the marriage of the hero and the hero having children) in that it suggests that $X(a_i)$ should be grouped in a clade with $X(c_{ii})$, rather than with $X(a_{ii})$. In a cladistic framework, competing scenarios like this are resolved by invoking the principle of parsimony, which states that scientific explanations should never be more complicated than is necessary. On that basis, the scenario in which $X(a_i)$ and $X(a_{ii})$ form a çlade is the more plausible scenario, since it requires a fewer number of evolutionary changes (i.e. character state transitions) than the alternative scenario linking $X(a_i)$ and $X(c_{ii})$: Whereas the former only requires four changes (one gain of the marriage, one gain of the hero and his wife having children, and two gains of the magic sword), the latter requires five (one gain of the magic sword, two gains of the marriage and two gains of the children); see Fig. 1.

While this example represents an extremely simple case, in most real-world datasets the number of characters and complexity of their distributions prohibit solution without the aid of a computer. Specialised bioinformatic software, such as

A.

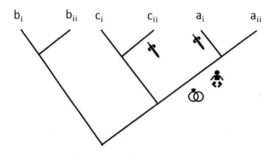

B.

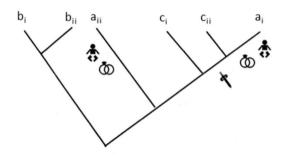

Fig. 1 Example of cladistic reconstruction. (**a**) and (**b**) present two alternative phylogenetic hypotheses for the distributions of three motifs: the hero's wedding (*rings symbol*), the hero has children (*baby symbol*) and the hero has a magic sword (*sword symbol*). *Taxon labels* representing different variants of a tale are given at the tips of the trees. Hypothesis (**a**) requires the sword to evolve twice to explain its presence in taxa a_i and c_{ii}. Hypothesis (**b**) requires the marriage and the children to both evolve twice to explain their presence in a_i and a_{ii}. Under the principle of parsimony, Hypothesis (**a**) is preferred (see text for details)

the popular *PAUP 4.0** (Swofford 1998) program, enable the researcher to explore billions of arrangements in order to determine the phylogenetic tree that minimizes the number of evolutionary changes required to explain shared character states. Even then, there is no guarantee that the most parsimonious tree will be an accurate model for the descent history of a tale. First, the usefulness of the outputs of a cladistic analysis relies on the quality of the input data, and in particular on the validity of the character data. How have the motifs been defined? Is the coding scheme rigorous and objective? Has it been applied consistently?

Secondly, it is important to evaluate the robustness of the relationships indicated by the most parsimonious tree. For example, if a particular clade is only supported by one or two more characters than an alternative (slightly less parsimonious) arrangement, then the hypothesized exclusive common ancestry of its members needs to be treated with caution. For this reason, researchers usually employ various statistics to assess how well the phylogenetic relationships returned by a cladistic analysis are supported by the data. These include the Consistency Index (CI) and Retention Index (RI), both of which quantify the degree to which the distributions of shared characters fit a given tree (Farris 1989; Kitching et al. 1998). The CI measures the number of character state changes required by the most parsimonious tree scaled against the sum of the minimum possible number of changes for each character (whereby each state would only have to evolve once). A CI of 1 indicates a perfect fit between the data and the cladogram, while values approach 0 with increasing levels of homoplasy. The RI measures the number of character-state changes required by a tree relative to the number that would be required if every taxon had evolved independently. Like the CI, the RI is scaled from 0 to 1, with higher values indicating a better fit between the tree and the data. Although both measures are widely used, the RI has a crucial advantage in that, unlike the CI, it is not sensitive to the number of characters or taxa, and thus can be useful for comparing phylogenetic signals across different datasets and trees.

A third method that is used to assess the robustness of cladistic reconstructions is bootstrapping (Felsenstein 1985). Bootstrapping involves generating a large number (e.g. 100,000) of "pseudo-datasets" by randomly sampling with replacement from the original character matrix. The pseudo-datasets are then subjected to cladistic analysis. Support for the clades returned by the original analysis is then estimated by calculating the frequency with which they occur in the most parsimonious trees obtained from the pseudo-datasets. Bootstrapping is thus particularly useful for identifying which relationships are the most robust and which ones are the weakest.

Although cladistic analysis is powerful and relatively easy to implement, there are some important limitations to the approach. In particular, the assumption of parsimony may be overly simplistic in many cases, such as when there is considerable variance in rates of evolution in different traits and/or different lineages. In a folklore context, it is certainly conceivable that some motifs might be more likely to be gained or lost than others. For example, folklorists (e.g. Thompson 1977) have suggested that motifs related to events in a story are more stable than motifs related to characterization (e.g. the gender, species, etc. of the protagonists), and that storytellers are more likely to alter the beginning and end of a tale than the core middle section of a narrative. In these instances, a "less parsimonious" reconstruction that allows some motifs (e.g. characterization, episodes that occur in the beginning or end of a story) to switch between states more freely might be more accurate than one which minimizes the overall number of character changes. To return to our earlier example, perhaps the wedding of the hero and him having children, both of which occur at the end of tale X, are more likely to evolve multiple times than the motif of the magic sword. In that case, the weight of the evidence in favor of clade $[X(a_i)$ and $X(a_{ii})]$ over clade $[X(a_i)$ and $X(c_{ii})]$ might shift in the other direction.

Similarly, since lineages do not always split at an even rate, we might expect some branches to be significantly longer than others, and want to allow more evolutionary change to occur on longer branches (e.g. multiple gains and losses of the same character state). This is especially important when dealing with international types that have a wide but patchy geographical distribution, as is often the case (Uther 2004). For example, if a tale has been extensively recorded in western Europe and East Asia, but not in intervening regions, we would expect the branch separating, say, French and Italian variants to be far shorter than the branch separating French and Italian variants from Chinese variants, and for the amount of mutation in motifs to vary accordingly.

Bayesian phylogenetic inference (Huelsenbeck et al. 2001; Matthews et al. 2011) provides an alternative method to cladistics that is better able to deal with these issues (Fig. 2). Bayesian inference proceeds by calculating the likelihood of the data (i.e. the chances of obtaining the observed distribution of character states) given an initial, randomly chosen, tree topology, a set of branch lengths and model of character evolution. The state of each parameter is then modified (i.e. clades are resorted, branches get lengthened/shortened, variance in rates of character change is increased/decreased) and the likelihood of the data is recalculated. This process is then repeated hundreds of thousands of times using an iterative technique known as Markov Chain Monte Carlo (MCMC) analysis. In this analysis, moves that improve the likelihood of the data are always accepted, while those that do not are usually rejected—although some may occasionally be accepted within a certain threshold. This is because the searching for the best parameter states (i.e. the most likely trees) is similar to walking through a mountain range; to find the highest peak sometimes you need to go down in order to find a route that leads higher up. However, in this case, the ground is constantly shifting below one's feet as each parameter gets adjusted simultaneously. Thus, the tree that seems to be the best one (i.e. maximizes the likelihood of the distribution of character states) under one set of conditions may turn out to be sub-optimal

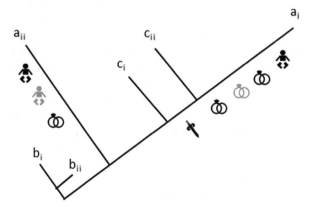

Fig. 2 Example of Bayesian inference. Bayesian inference allows for rates of evolution to vary for different motifs and branches of the tree. Here, the hero's marriage and the hero's children are represented as fast evolving traits that are each gained, lost and regained in separate lineages. The length of the branches corresponds to the amount of evolutionary change that is expected to occur in each lineage (with more evolution occurring on longer branches)

when the branch lengths or variance in rates of character change are slightly adjusted (whilst at the same time, the best values for these parameters will vary with different tree topologies). Bayesian phylogenetic inference integrates the uncertainty associated with alternative evolutionary scenarios by sampling trees at regular intervals in the MCMC approach to compile a "posterior distribution of trees". Since the analysis usually favors moves that increase the likelihood of the data, it "revisits" higher peaks in the likelihood landscape more frequently than the lower peaks, meaning that trees with higher probabilities are sampled more often than ones with lower probabilities. Once the posterior distribution of trees has been compiled, phylogenetic relationships among tale variants can be represented by a consensus tree that shows the "posterior probabilities" of individual clades, which correspond to the percentage of posterior trees which they occurred in. The latter provides a useful indication of the robustness of these relationships under a range of plausible evolutionary models, rather than just a single optimality criterion like parsimony. There are several software programs for carrying out Bayesian phylogenetic analysis, such as the popular open source package MrBayes (Ronquist et al. 2012).

While cladistics and Bayesian phylogenetic inference employ different esti-mation procedures, both are primarily concerned with reconstructing "vertical" lineages of inheritance, i.e. the descent of tale variants from common ancestral traditions. However, it may often be desirable to explore other patterns of resem-blance that result from processes such as independent evolution or the exchange of motifs between phylogenetically distinct lineages, which cannot be captured by a tree model. For example, in our hypothetical tale X we have discussed conflicting distributions of motifs like the hero's marriage and the magic sword as a problem that has to be resolved in order to establish the correct phylogenetic relationships of $X(a_i)$, $X(a_{ii})$ and $X(c_{ii})$. However, whichever tree we choose suppresses information about the relationships among these variants and therefore does not fully represent their evolutionary histories. This problem can be addressed using "phylogenetic network methods", which do not employ a bifurcating tree model. One of the most widely used methods of this kind is NeighborNet, an algorithm implemented in the program SplitsTree (Huson and Bryant 2006).

NeighborNet uses a character matrix to calculate pairwise distances between taxa, which represent the average number of mutations per character that separate one taxon (e.g. tale variant) from another. The analysis then progressively partitions the taxa into a series of "splits", analogous to clades, in which each taxon is paired with its nearest neighbor. When two pairs overlap (i.e. where the same taxon is represented twice), they are agglomerated to create two composite taxa. For example, if $X(a_i)$ forms a pair with $X(a_{ii})$ and another pair with $X(c_i)$ then each pair is agglomerated to form $X(a_i)/X(a_{ii})$ and $X(a_i)/X(c_{ii})$. The distance of a composite taxon to all the remaining taxa is averaged from the two original taxa, and further splits are calculated. This process is repeated until a complete series of splits for the data have been obtained and no further agglomerations are possible. A key feature of the technique is that it allows taxa to be split in multiple, potentially conflicting, ways. These relationships are displayed in the form of a "splits graph",

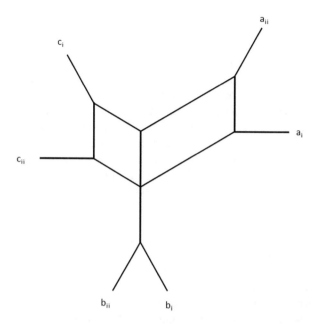

Fig. 3 Example of a NeighborNet: A NeighborNet "splits graph" can display conflicting signals in the data, represented by the *box-like shapes*. Each pair of parallel edges in the boxes represents a partition among the taxa, in this case grouping a_{ii} and a_i, and a_{ii} and c_i. The length of each pair of edges corresponds to the support for the partition in the data, with *shorter edges* representing smaller distances (i.e. number of mutations) separating taxa

which shows groupings in the data (represented by parallel edges) and distances separating them (which are proportional to the lengths of the parallel edges). Where the splits are highly consistent, the diagram will resemble a branching tree-like structure. Incompatible splits, on the other hand, produce box-like structures that lend a more latticed appearance to the network (Fig. 3). Although the complexity of these diagrams can make it difficult to read the evolutionary history of a tale, they provide a useful visualization of conflicting patterns in a dataset that complement cladistic and Bayesian trees.

Empirical Studies

Having outlined the basic principles of applying phylogenetic analysis to oral narratives and discussed some of the main techniques, we turn now to two case studies that demonstrate the approach.

Case Study 1: Little Red Riding Hood (ATU 333)

Our first case study concerns one of the most famous and controversial international tales in the folklore literature—Little Red Riding Hood. This is classified as ATU 333 in the Aarne-Thompson-Uther (ATU) Index of International Tale Types. The central plot of the tale concerns a young girl who visits her grandmother's house, where she is eaten by a wolf disguised as the old woman. Most modern versions of the tale can be traced back to 1697, when the first classic version of the story, *Le Petit Chaperon Rouge*, was published by the French author Charles Perrault (1697). While some researchers (Dundes 1989) have suggested that Perrault may have invented the story, it is generally believed that he probably based it on traditional sources (Zipes 1993). In particular, it is thought that *Le Petit Chaperon Rouge* was derived from a folktale known as 'The Story of Grandmother', versions of which survived into the nineteenth and twentieth centuries in the oral literatures of southeast France, the Alps and northern Italy (Dundes 1989; Zipes 1993). These variants of ATU 333 are often considerably more graphic and violent than the Little Red Riding Hood that many of us grew up with. In some versions for example, when the young girl arrives at the house she complains that she is hungry, so the wolf offers her a meal rustled up from her poor grandmother's remains (e.g. "tortellini" that are actually ears). Other important differences concern the resolution of the plot. In Perrault's tale the girl is eaten by the wolf and dies, while in later versions, such as Grimm's classic telling (Grimm 1884), she is rescued from the wolf's stomach by a passing hunter. However, in 'The Story of Grandmother', the young girl typically outwits the wolf/werewolf by tricking him into letting her go outside to urinate, whereupon she flees into the woods. Of course, since these oral tales were all collected long after the publication of Perrault's tale, there is no way of knowing for sure whether they are descended from a more ancient folktale tradition. However, there is some evidence to suggest that some of the most basic and characteristic motifs of Little Red Riding Hood have been in existence at least since medieval times. This is supported by an eleventh century Latin poem by Egbert of Liége (Ziolkowski 1992), which relates a local Walloon folktale in which a young girl, dressed in a red tunic, gets lost in the woods and is attacked by a wolf. The wolf takes her back to its home, where she manages to escape (thanks to the supernatural protection of her red tunic).

While all these tales are considered to be versions of the same type (ATU 333), Little Red Riding Hood is also thought to be related to another well-known international tale, ATU 123 'The Wolf and the Kids' (Dundes 1989). ATU 123 is popular throughout Europe and the Middle East, and can be traced back to an old Aesopic fable that was first recorded in the first century C.E. (Tehrani 2013). The story tells of a nanny goat who leaves her kids at home while she goes out to forage for food. In her absence a wolf comes to the door and tricks the kids by impersonating their mother. They let the wolf in and are devoured. When the nanny goat discovers what has happened she kills the wolf and rescues her kids by cutting them out of the wolf's stomach with her sharp horns (echoing the ending of some versions of Little Red Riding Hood). While there are clear similarities between these two tales, they are distinguished as separate international types by two main characteristics: (1) The victim in ATU 333 is a human girl, while in ATU 123 it is a litter of goat kids; (2) In ATU 333 the victim is attacked in her grandmother's house, whereas in ATU 123 the victims are attacked in their own house.

In the case of western folklore traditions, the definitions of ATU 123 and ATU 333 are relatively clear cut. However, in other cultural contexts the distinctions described above are much more problematic. For example, a story found in central and southern Africa and parts of the Caribbean tells of a young girl who is left home alone by her mother or guardian. An ogre comes to her house disguised as her mother and she lets the monster in, and is eaten (in some versions of this tale, she is subsequently cut out of the ogre's belly alive). Thus, in terms of the two key traits that distinguish ATU 333 from ATU 123, this tale would be classified as ATU 333 on the basis of it featuring a human girl as the victim, but as ATU 123 on the basis of the attack occurring in her own home. An even more complex case is presented by a popular Chinese tale 'The Tiger Grandmother', other variants of which have been recorded throughout East Asia. This story tells of a group of siblings who are left at home by their mother. A tiger or monster comes to the house disguised as their grandmother, and tricks the siblings into letting it in. The tiger gets into bed with the children and during the night the older children are woken by the sound of the monster crunching on the bones of their youngest sibling. Eventually, they manage to escape by tricking the monster into letting them outside to go to the toilet. Again, this tale exhibits an array of features linking it to both ATU 333 and ATU 123. The victims are human, like the girl in Little Red Riding Hood/Story of Grandmother, but there are several of them, like the goat kids in ATU 123. The victims are attacked in their own home, as per The Wolf and the Kids, but in a manner that is highly reminiscent of Little Red Riding Hood, who is also attacked in bed. Moreover, the children in Tiger Grandmother employ the same ruse as the heroine in The Story of Grandmother to escape the monster.

The difficulty of classifying these African and East Asian stories as either ATU 123 or as ATU 333 epitomizes the problems with current folktale taxonomies mentioned at the beginning of this chapter. Definitions of international types based on just a few motifs are unlikely to prove adequate for establishing relationships among stories distributed in many different regions of the world, particularly when the motifs in question are derived from specifically western tale traditions. To overcome these problems, Tehrani (2013) used a phylogenetic approach to analyze a much larger set of motifs exhibited by these tales to address (a) whether they can be divided into coherent and distinct groups corresponding to international types ATU 123 and ATU 333; and (b) if so, which type the Asian and African tales belong to.

Tehrani's analysis focused on a sample of 58 versions of ATU 333 and ATU 123 from 33 populations (including East Asian and African groups). Based on detailed comparisons of the content of these tales, 72 motifs were defined. They included features such as character of the protagonist (single child versus group of siblings; male versus female), the character of the villain (wolf, ogre, tiger, etc.), the tricks used by the villain to deceive the victim (false voice, disguised paws, etc.), whether the victim is devoured, escapes or is rescued, and so on (see Tehrani 2013, for a full list). The states of these motifs in each variant were recorded in a matrix, which was then analyzed using cladistic methods, Bayesian inference and NeighborNet, as described in the previous section. The results of these analyses are shown in Fig. 4.

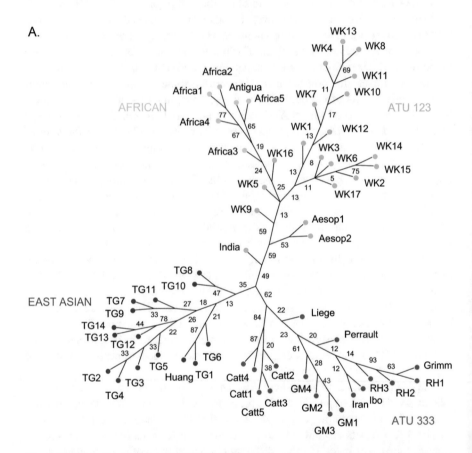

Fig. 4 Phylogenetic analyses of ATU 333/123. (**a**) Results of a cladistic analysis. Numbers beside the internal branches represent bootstrap support percentages for the corresponding clade. (**b**) Results of a Bayesian analysis. Numbers beside the internal branches represent posterior probabilities for the corresponding clade. (**c**) NeighborNet splits graph obtained from the dataset. The labels on the leaf nodes refer to the tales used in the analysis. Many of these are variants of the same story and are labeled with abbreviations (*WK* Wolf and the Kids; *RH* Little Red Riding Hood; *GM* Story of Grandmother; *Catt* Catterinetta, an Italian ecotype of Little Red Riding Hood; *Africa* African ogre tale; *TG* Tiger Grandmother). Historically notable versions are given individual names (e.g. Perrault; Grimm, Aesop, etc.). Major groups in the data are color-coded (ATU 123 *green*, ATU 333 *red*, East Asian tales *purple*, African tales *blue*)

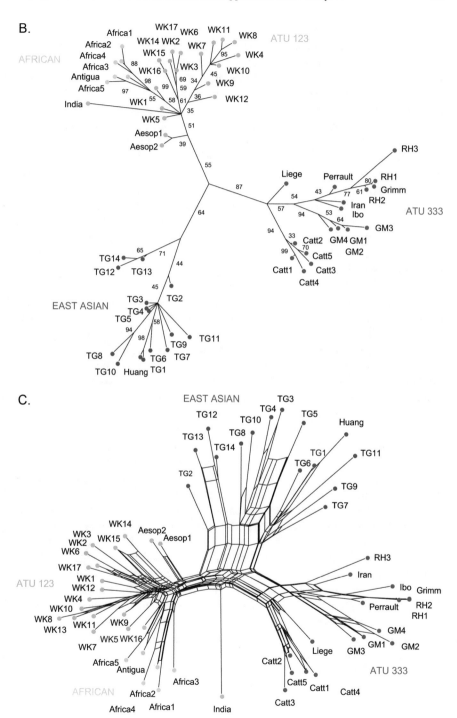

Fig. 4 (continued)

Overall, the results of the analyses suggest that it is possible to discern a reasonably strong historical signature in the distribution of motifs among the variants. The tree returned by the cladistic analysis had a Retention Index of 0.72, which suggests that the majority of similarities among the tales are consistent with a branching evolutionary process (Collard et al. 2006; Nunn et al. 2010). Moreover, all three analyses were able to recover relationships among specific tales that are known from the historical record: In accordance with the chronological record, relationships within the ATU 333 group indicate that Little Red Riding Hood and the Story of Grandmother are descended from a common ancestor that existed more recently than the last ancestor they share with the eleventh century Liège poem. The position of the Grimms' version of Little Red Riding Hood supports historiographical evidence that it is directly descended from Perrault's earlier tale (via a literate informant of French Huguenot extraction). The results of the analyses also concur with the literary record on The Wolf and the Kids, which suggests the tale evolved from an Aesopic fable which was first recorded around 400 AD. All three analyses indicate that Aesopic versions of the tale diverged at an early point in the history of the lineage.

Close inspection of the trees and network reveal some subtle differences that are informative about the methods. For example, the Bayesian analysis suggests that a version of The Wolf and the Kids from India evolved after the Aesopic fable, whereas the cladistic analysis and NeighborNet splits graph place it outside the group containing the other ATU 123 variants. This probably reflects the fact that the Indian tale is an extremely divergent version of ATU 123, which the Bayesian approach is better placed to handle because it explicitly models variance in rates of evolution and branch lengths, whereas the other two methods do not.

However, despite these minor differences, all three analyses returned the same major groupings. They each supported a split between the established international tale types ATU 123 and ATU 333, comprising the western tales of The Wolf and the Kids and Little Red Riding Hood/The Story of Grandmother respectively. The analyses all grouped the African tale with variants of The Wolf and the Kids. This suggests that this story can be considered an African ecotype of ATU 123, and that its resemblances to Little Red Riding Hood are due to independent evolution (or perhaps borrowing). The East Asian tale 'Tiger Grandmother' on the other hand did not group with either ATU 333 or ATU 123, but formed a separate cluster and therefore cannot be classified as either type. Instead, there are three possible explanations for the relationship of this tale to The Wolf and the Kids and Little Red Riding Hood. The first is that it evolved entirely independently. Although this cannot be ruled out, the number of shared motifs and the way that they are functionally integrated into such similar plots suggests that there is much more likely to be some kind of historical connection—especially given the extensive exchange of stories (as well as other cultural traits) between East and West (e.g. Haar 2006). The second possibility is that the East Asian tales represent a separate lineage that is descended from the same archetype tale that gave rise to ATU 333 and ATU 123 (e.g. Dundes 1989; Haar 2006). Under this scenario, the traits shared by The Tiger Grandmother and The Wolf and the Kids on the one hand, and those shared by The

Tiger Grandmother and Little Red Riding Hood on the other would be expected to be homologous—ancient inheritances from their last common ancestor.

To test this hypothesis, Tehrani (2013) carried out a procedure known as "ancestral state reconstruction", which involves mapping shared traits on a tree to establish patterns of homology and homoplasy. He found that most of the key resemblances between the Tiger Grandmother and the other tales were homoplastic—suggesting they were not inherited from an original archetype tale. The third possibility is that The Tiger Grandmother evolved by borrowing and blending elements of both ATU 123 and ATU 333 (and probably local folktales) to create a new "hybrid" tale. This hypothesis would explain the homoplastic similarities between The Tiger Grandmother and Little Red Riding Hood and The Wolf and the Kids. It is also consistent with the structure of the NeighborNet splits graph, which suggests a particularly high concentration of conflicting relationships surrounding the Tiger Grandmother. Consequently, the "hybridization" hypothesis seems to be the most plausible explanation for the relationships between The Tiger Grandmother and western folktales. In sum, these analyses demonstrate how phylogenetic methods can be used to address long-standing controversies surrounding the classification of non-western tales, and shed new light on the historical evolution of international types.

Case Study 2: Polyphemus (ATU 1137)

Our second case study concerns another widespread folktale, which has been documented since ancient times: ATU 1137 'Polyphemus'. The title of this tale type is drawn from the earliest and most famous version of the story, which occurs at the beginning of Homer's *Odyssey* (probably composed near the end of the eighth century BC). It recounts how Odysseus landed on an island with his men and found refuge in a large cave. After some time, the cave's inhabitant, a Cyclops named Polyphemus, returned with his flock of giant sheep and placed a huge rock across the mouth of the cave to seal the entrance. Finding Odysseus and his men, Polyphemus promptly grabbed and ate two of them, and then fell asleep. In the morning, the Cyclops rolled the massive stone aside, let his sheep out to pasture, and closed the entrance again. That evening, Odysseus offered the Cyclops wine and said that his name was "Nobody". Polyphemus got drunk and fell asleep. At night, Odysseus and his companions blinded the ogre with a sharp point hardened in the fire. Polyphemus called his fellow Cyclopes for help, telling them that it was all the fault of "nobody". The ogres returned home, thinking that Polyphemus was being afflicted by divine power. When Polyphemus awoke and led the sheep out of the cave, he stood by the cave's entrance and felt the back of each animal to prevent the Greeks' escape. However, Odysseus and his men tied themselves to the undersides of the animals to avoid detection and got away.

A remarkably similar plot can be found in North America. For instance, the Jicarilla Apaches (Opler 1938: 256–260) tell of two young men who are turned into puppies and found by the children of Raven, who take them home. The father Raven was suspicious, so kindled a fire in front of them. The first puppy blinked and was driven off. Raven was finally persuaded that the second, which didn't blink, was a real dog. The puppy observed that Raven opened a door in a cliff when he wanted to kill some buffalo. At night, when they had all gone to bed, the dog turned back into a person and opened the door to let the buffalo out. Raven fetched an arrow to kill the man, but he escaped by hiding in the last buffalo's anus to escape.

As these brief summaries show, there are close similarities between the plots of these Amerindian and European tales. Each time, a person gets into the homestead of a master of animals or of a monstrous shepherd, often with a false identity (e.g. "Nobody", a puppy), to search for food; and the homestead is used to retain or keep animals. The door is often a huge stone. The host wants to kill the hero and waits for him to pass by, but the hero escapes by hiding in or under an animal that is going out. Fire is usually used to destroy the European giant and to burn the Amerindian Raven, or the enemy of Raven. It seems highly unlikely that such complex sets of traits and similar structures could have evolved independently. Two explanations arise: First, the narrative was carried to the Americas by European colonists during or after the late medieval period; Second, the story spread across Eurasia and North America when a former land bridge joined present day Alaska and eastern Siberia during the Pleistocene ice age (Berezkin 2007: 83–84 and Chapter "Peopling of the New World from Data on Distributions of Folklore Motifs").

To test these hypotheses, d'Huy carried out a series of phylogenetic analyses (d'Huy 2012a, 2013a, 2015), the last one focusing on 190 traits exhibited by 56 versions of Polyphemus sampled from a wide range of cultures. The relationships among the versions were reconstructed using cladistic analysis, Bayesian inference and NeighborNet. The trees were rooted using a method known as "midpoint rooting", which places the root at the mid-point of the longest distance between two taxa in the tree, which in this case were between the Ojibwa version (a group of Native Americans in North America) and Valais version (one of the 26 cantons of Switzerland). The historical background behind the branching of the Ojibwa and Valais examples is unclear. It is likely that the Valais version, which includes a lord of wild animals similar to Amerindian versions, exhibits the most archaic features. That would be in agreement with Burkert's statement (1979) that the Cyclops was primitively a lord of animals. However, it remains unclear why the primitive version would be maintained only in the traditional area of Valais where it is attested in only one variant. We could suppose that the local evolution of this tale shaped it like the (Palaeolithic) proto-form, which explains its place in our analysis, and not historical continuity from such an era before the domestication of livestock.

All the methods and datasets used show a remarkably consistent pattern, wherein versions from geographically neighbouring populations tend to form sister clades). We note that the current geographical distribution of the Polyphemus' tales is consistent with the diffusion of the haplogroup X2 in both North America and Europe. This mitochondrial DNA haplogroup should have moved from Beringia directly to the North American regions around 18000 BP, before the Bering land bridge disappeared (Reidla et al. 2003). The co-occurrence of the versions and of the haplogroup is remarkable and provides intriguing evidence about the antiquity of the tale. Accordingly, the first versions at the base of the tree (after the Valais version for the European part) are Mediterranean and near the American Great Lakes, both of these areas being considered as refuges during the last Glacial Maximum.

If Polyphemus is indeed a Paleolithic tale, then one can expect the transmission of motifs to be very stable. To test this, the Retention Index (RI—see above) was calculated for trees derived from different sets of motifs. The RIs oscillate between 0.57 and 0.79 (d'Huy 2013a, 2014a). By way of comparison, a comparative study by Collard et al. (2006) found the average RI returned by phylogenies of biological datasets (which can be assumed to be structured by branching descent with modification) was 0.61, while other cultural datasets returned an average RI of 0.59. The RI scores associated with the Polyphemus trees thus indicate that, as expected, the motifs associated with different versions of Polyphemus have been highly conserved.

The phylogenetic stability of motifs makes it possible to trace their transmission histories and make inferences about the contents of ancestral variants (d'Huy 2012b) using the same techniques of ancestral state reconstruction as Tehrani (2013) used in his study of *Little Red Riding Hood*, discussed above. One such method applies the principle of parsimony, which selects ancestral states (e.g. the presence or absence of a particular feature) that minimise the number of character state changes (i.e. mutations) on each branch leading to the tips of the tree (i.e. the existing tales). Other methods, such as maximum likelihood or Bayesian inference, take a probabilistic approach which quantify the likelihood of each character state given their distributions on the tips of the tree, an estimated rate of change and set of branch lengths (so that more change is permitted on longer branches than short ones). According to both of these methods, from two datasets of traits and different versions, the original *ur*-tale could be reconstructed as follows (d'Huy 2013a, 2015): "A human hunter enters a monster's house, which is a hut, a house or something similar. The hero is uninvited, and comes with the express purpose of stealing something, and does not know who he will meet. The monster possesses a herd of wild animals, which are locked up. He traps the man and his own animals with an immovable or a large door. Then he waits for the man near the entrance to kill him and checks the animals that go out. To escape, the hero clings to a living animal. In this story, vengeance occurs, connected with fire". Of course, ancestral state reconstruction does not produce a fully-fledged telling of the *ur*-tale.

Rather it allows us to infer the basic building blocks from which subsequent variants evolved.

This reconstruction is potentially supported by archaeological data. For instance, an illustration from the Magdalenian cave drawings, the Trois-Frères, may be interpreted as from the Polyphemus tale. The potential case includes a scene within a dense, superimposed and complex representation of a bison herd. It depicts a bison-man with a bow in his hand who observes one of the animals which the Magdalenian cave drawings—if so interpreted—has a human thigh and a very detailed, large anus or vulva. This peculiar image can be compared to the Amerindian versions of the Polyphemus tale, in which the hero often hides inside an animal itself by entering through its anus to escape from the supernatural guardian of the herd, who controls the beasts from his dwelling (d'Huy 2015).

The diffusion of the tale in Europe is quite complex, and different versions of the story may have been 'seeded' through Europe again and again, superseding one another and receding in the wake of history. Accordingly, there is low relationship between geographic distance and similarity between versions (7 % of the variance explained in the Amerindian data and 0.8 % in the Eurasian data; see d'Huy 2015), each new version taking the place of older versions, breaking the continuity of linear diffusion and making it difficult to correlate the earliest, palaeolithic reconstructable version of the tale with any particular geographical space. However, it is possible to attempt to correlate the trees with a model of historical diffusion based on prehistoric population dispersals. For instance, in Europe, the palaeolithic populations may have migrated toward the South (Greece, Syria, Abaze) during the Last Glacial Maximum. The tale was elaborated here and a new version appears, where the monster possesses a shelter and the animals were sheep. This version may go back to the domestication of sheep, which is estimated to fall between 9 and 11,000 years ago in Mesopotamia. If the new Polyphemus' tale type was linked to the early stages of animal domestication, it may have been disseminated through successive migrations from the Mediterranean area across millennia, replacing the ancient story.

As with the possible Palaeolithic origin of versions of Polyphemus, tales that are passed on from generation to generation within the same community tend to be quite stable and accumulate mutations gradually. When a tale travels from one group to another, or is transmitted by a human migration with a very small group, on the other hand, it often undergoes a period of rapid change as its content is adapted to suit a new cultural environment. Such events leave strong phylogenetic signatures that are described by the term "punctuated equilibrium", characterised by sudden bursts of speciation, followed by prolonged periods of relatively gradual change and evolutionary stasis (Eldredge and Gould 1972; Gould and Eldredge 1977). Statistical techniques for detecting punctuated equilibrium (obtained by measuring the length of a tree against the number of branching events) thus provide a useful means to assess the impact of ecotypification on oral traditions. Analyses of Polyphemus (d'Huy 2013a, 2015) suggest that these tales exhibit a highly punctuated pattern of development that is consistent with models of ecotypification. The punctuation effect for this folktale contributes between 60 and 83 % to its evolution, a result close to what was obtained for another tale-type: the Cosmic Hunt (84 %; d'Huy 2013b). The remaining variation in path could be explained by independent gradual effects. This effect is stronger than the punctuation effect in biological species (22 %; Pagel et al. 2006) or in languages (10–33 %; Atkinson et al. 2008). As the folklorist Etunimetön Frog succinctly puts it, "the evolution of tradition [occurs] in fits and starts" (Frog 2011: 91) and it is now possible to quantify this effect (Fig. 5).

A.

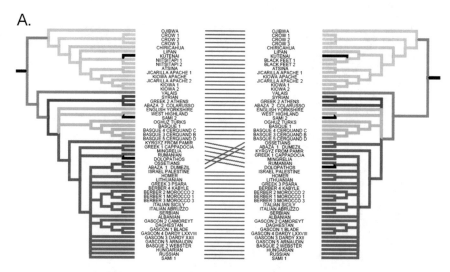

Fig. 5 Phylogenetic analyses of ATU 1137. (**a**) Tree under the maximum parsimony consensus and Bayesian maximum clade credibility tree. The *blue cluster* is the primitive Mediterranean group, where the versions with domestic animals have been elaborated. The *green clusters* include Basque, Oghuz Turks, Yorkshire and the West Highlands. It probably takes place after the diffusion of the tale in Europe from the Mediterranean area. During the first millennium BC, Celtic languages were spoken across much of Europe, including Great Britain, the Pyrenean area, the Black Sea and the Northern Balkan Peninsula. The Basque versions may be borrowings from the neighbouring Celtiberian (spoken in ancient times in the Iberian Penninsula) or Gaulish languages. The link between the Pyrenean area and Oghuz Turks could be explained by the Gallic invasion of the Balkans in 279 BC. The *red cluster* probably shows the pre-Homeric elaboration of the tale in the Mediterranean area. The *magenta cluster* may be link to the diffusion of the Odyseus version. However, this hypothesis could not be shown to be 'true', but as a possible but undemonstrable explanation. (**b**) NeighborNet graph of the Polyphemus variants. (**c**) The 'Petit Sorcier á l'Arc Musical' ['The Sorceror with the Musical Bow'] in the Cave of the Trois-Frères in Ariège, southwestern France, Magdalenian, may be the earliest pictographic representation of the Polyphemus tale (Breuil 1930: 262)

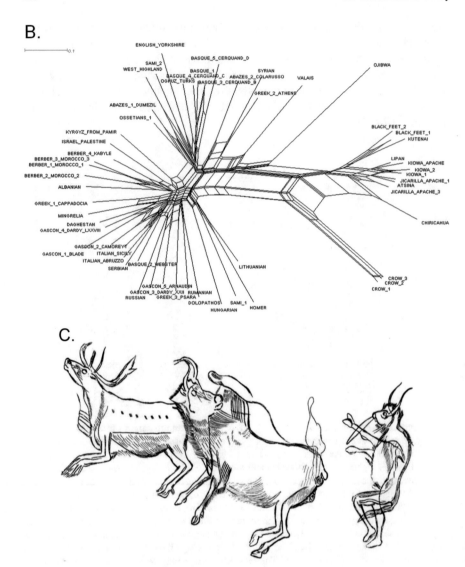

Fig. 5 (continued)

Future Directions

The field of folklore phylogenetics has made great strides in a relatively short period of time. The case studies we have reviewed here demonstrate that modern evolutionary theory and methods provide a new and powerful framework for addressing the kinds of questions that have long interested folklorists, whilst overcoming previously intractable problems associated with more traditional methods. Nevertheless, it remains a young field, with plenty of scope to mature. We therefore conclude our chapter by briefly pointing to some areas that we consider to be particularly fertile for further growth.

The first, and perhaps most important, issue concerns the definition of characters for analysis. In terms of how data are coded and prepared for analysis, folklore phylogenetics remains rather unreconstructed in its approach: like earlier generations of the historic-geographic school, the identification of motifs and traits is determined by an individual researcher. However much we may strive to be sensible and rigorous, there will always be a risk that our datasets (and hence, our results) might be shaped by our personal biases and idiosyncrasies. One solution might be to compare analyses based on multiple datasets to determine how robust inferred phylogenetic relationships are to different coding schema (d'Huy 2013a, b, c, d, 2014a, b). A more direct way of addressing the problem would be to develop machine-based methods for deriving motifs from texts based on specific linguistic signatures (d'Huy 2013c, 2014c, 2015). Failing that, we believe it would be useful to establish a set of common principles that would lead to replicable outcomes for any given set of tales.

A second question surrounding characters is whether there are any general rules governing the transformation of motifs. To what extent are such characters able to change independently of one another? Or are there certain structural relationships that link certain features together (e.g. Lévi-Strauss 1971), which our analyses should control for? Perhaps certain segments of a narrative are freer to change than others. For example, Aarne (cited in Thompson 1977:436) suggested that the beginning and end of a tale are more likely to be improvised than the core middle section, where changes would be likely to undermine the coherence of the narrative. Such a hypothesis would certainly merit testing in a phylogenetic analysis, and may have important implications for reconstructing the deepest layers of a tradition. Similarly, there are potentially rich links to be made between folklore phylogenetics and research in cognitive psychology and cultural transmission theory (e.g. Sperber 1996; Mesoudi 2011; Chapter "Cognitive and Network Constraints in Real Life and Literature") about the ways in which information is encoded and recalled in human brains. This work suggests that certain kinds of representations—for example ones that elicit an emotional response or have counterintuitive properties—may be intrinsically more attractive and memorable than others, and thus be better preserved in narrative transmission (d'Huy 2013e; Stubbersfield and Tehrani 2013; Stubbersfield et al. 2015). Folklore phylogenies provide an outstanding natural laboratory for testing these hypotheses.

Another area that merits further attention concerns the relationships between narrative lineages and population histories (see Berezkin's contribution to Chapter "Peopling of the New World from Data on Distributions of Folklore Motifs"). As the case studies reviewed above show, phylogenetic connections among tales told in different cultures are likely to reflect complex processes of dispersal, migration and interaction. At present, the characteristic signatures of these processes and their relative importance in the evolution of oral traditions' remain poorly understood. However, some studies (e.g. Ross et al. 2013; d'Huy 2012c, 2013b, e; d'Huy and Dupanloup 2015) have already begun to address this problem by exploring the extent to which variation in international tale types is structured by genetic, linguistic and geographic relationships among populations. We anticipate that further studies in this area will generate significant insights into the processes responsible for generating and maintaining international tale traditions.

Last of all, while this chapter has stressed the fundamental similarities between reconstructing the evolutionary histories of organisms and folklore traditions, we do not claim that the methods developed by biologists are perfect, or that they should be adopted uncritically. The applicability of any method depends on the specific nature of the question that is being asked, and future progress in folklore phylogenetics will probably require the development of new methods tailor-made to address specific problems in the study of narrative traditions that do not have direct counterparts in evolutionary biology (e.g. Tehrani et al. 2015). Such efforts will need to go beyond simply importing ready-made tools from bioinformatics, and require an even more ambitious interdisciplinary enterprise bringing together folklorists, literary scholars, biologists, mathematicians and computer scientists.

References

Atkinson, Q. D., Meade, A., Venditti, C., Greenhill, S., & Pagel, M. (2008). Languages evolve in punctional bursts. *Science, 319*, 588.

Berezkin, Y. (2007). Dwarfs and cranes: Baltic-Finnish mythologies in Eurasian and American perspective (70 years after Yrjö Toivonen). *Folklore, 36*, 75–96.

Breuil, H. (1930). Un dessin de la grotte des Trois frères (Montesquieu-Avantès) Ariège. *In Comptesrendus des séances de l'année 1930: Académie des inscriptions et belles-lettres, 74*(3), 261–264.

Burkert, W. (1979). *Structure and history in Greek myth and ritual*. Berkeley: University of California Press.

Collard, M., Shennan, S., & Tehrani, J. (2006). Branching, blending, and the evolution of cultural similarities and differences among human populations. *Evolution and Human Behavior, 27*, 169–184.

d'Huy, J. (2012a). Le Conte-Type de Polyphème. *Mythologie Française, 248*, 47–59.

d'Huy, J. (2012b). Un ours dans les étoiles: recherche phylogénétique sur un mythe préhistorique. *Préhistoire du sud-ouest, 20*(1), 91–106.

d'Huy, J. (2012c). Le motif de Pygmalion: origine afrasienne et diffusion en Afrique. *Sahara, 23*, 49–58.

d'Huy, J. (2013a). Polyphemus (Aa. Th. 1137): A phylogenetic reconstruction of a prehistoric tale. *Nouvelle Mythologie Comparée/New Comparative Mythology, 1*, 3–18.

d'Huy, J. (2013b). A cosmic hunt in the Berber sky: A phylogenetic reconstruction of Palaeolithic mythology. *Les Cahiers de l'AARS, 16*, 93–106.

d'Huy, J. (2013c). Le motif du dragon serait paléolithique: mythologie et archéologie. *Préhistoire du sud-ouest, 21*(2), 195–215.

d'Huy, J. (2013d). Il y a plus de 2000 ans, le mythe de Pygmalion existait en Afrique du nord. *Préhistoires Méditerranéennes*, 4.

d'Huy, J. (2013d). Rôle de l'émotion dans la mémorisation des contes et des mythes. *Mythologie Française, 253*, 21–23.

d'Huy, J. (2014a). Mythologie et statistique: reconstructions, évolution et origines paléolithiques du combat contre le dragon. *Mythologie Française, 256*, 17–23.

d'Huy, J. (2014b). Une méthode simple pour reconstruire une mythologie préhistorique (à propos de serpents mythiques sahariens). *Les Cahiers de l'AARS, 17*, 95–104.

d'Huy, J. (2014c). Motifs and folktales: A new statistical approach. *The Retrospective Methods Network Newsletter, 8*, 13–29.

d'Huy, J. (2015). Polyphemus, a Palaeolithic tale? *The Retrospective Methods Network Newsletter, 9*(Winter 2014–2015), 43–64.

d'Huy, J., & Dupanloup, I. (2015). D'Afrique en Amérique : la bonne et la méchante fille (ATU 480). *Nouvelle Mythologie Comparée/New Comparative Mythology* 2 (2014–2015), *online*.

Dundes, A. (1989). *Little red riding hood: A casebook*. Madison: University of Wisconsin Press.

Eldredge, N., & Gould, S. J. (1972). Punctuated equilibria: An alternative to phyletic gradualism. In T. J. M. Schopf (Ed.), *Models in Paleobiology* (pp. 82–115). San Francisco: Freeman Cooper.

Farris, J. S. (1989). The retention index and homoplasy excess. *Systematic Zoology, 38*, 406–407.

Felsenstein, J. (1985). Confidence limits on phylogenies: An approach using the bootstrap. *Evolution, 39*, 783–791.

Frog, E. (2011). Circum-Baltic mythology? The strange case of the theft of the thunder-instrument (ATU 1148b). *Archaeologia Baltica, 15*, 78–98.

Gennep, A. (1909). *Religions, mœurs et légendes 2* (Vol. 2). Paris: Mercvre de France.

Gould, S. J., & Eldredge, N. (1977). Punctuated equilibria: The tempo and mode of evolution reconsidered. *Paleobiology, 3*(2), 115–151.

Gray, R. D., Bryant, D., & Greenhill, S. J. (2010). On the shape and fabric of human history. *Philosophical Transactions of the Royal Society, B, 365*, 3923–3933.

Grimm, J., & Grimm, W. (1884). *Children's and household tales*. London: George Bell.

Haar, B. J. (2006). *Telling stories: Witchcraft and scapegoating in Chinese History*. Leiden and Boston: Brill.

Hafstein, V. T. (2001). Biological metaphors in folklore theory: An essay in the history of ideas. *Arv, 57*, 7–32.

Howe, C. J., Barbrook, A. C., Spencer, M., Robinson, P., Bordalejo, B., et al. (2001). Manuscript evolution. *Trends in Genetics, 17*, 147–152.

Huelsenbeck, J. P., Ronquist, F., Nielsen, R., & Bollback, J. P. (2001). Bayesian inference of phylogeny and its impact on evolutionary biology. *Science, 294*, 2310–2314.

Huson, D. H., & Bryant, D. (2006). Application of phylogenetic networks in evolutionary studies. *Molecular Biological Evolution, 23*, 254–267.

Kitching, I. J., Forey, P. L., Humphries, C. J., & Williams, D. (1998). *Cladistics: The theory and practice of parsimony analysis*. Oxford: Oxford University Press.

Lévi-Strauss, C. (1971). *Mythologique 4 L'Homme nu*. Paris: Plon.

Matthews, L. J., Tehrani, J. J., Jordan, F. M., Collard, M., & Nunn, C. L. (2011). Testing for divergent transmission histories among cultural characters: A study using Bayesian Phylogenetic methods and Iranian tribal textile data. *PLoS One, 6*, e14810.

Mesoudi, A. (2011). *Cultural evolution: How Darwinian theory can explain human culture and synthesize the social sciences*. Chicago: University of Chicago Press.

Nunn, C. L., Arnold, C., Matthews, L., & Mulder, M. B. (2010). Simulating trait evolution for cross-cultural comparison. *Philosophical Transactions of the Royal Society, B, 365*, 3807–3819.

Opler, M. E. (1938). *Myths and tales of the Jicarilla apache Indians* (Vol. 31). New York: Memoires of the American Folklore Society.

Pagel, M., Venditti, C., & Meade, A. (2006). Large punctuational contribution of speciation to evolutionary divergence at the molecular level. *Science, 314*(6), 119–121.

Perrault, C. (1697). *Histoires ou Contes du temps passé.. Avec des Moralitez*. Paris: Claude Barbin.

Propp, V. (1968). *Morphology of the folktale* (2nd ed.). Austin: University of Texas Press.

Reidla, M., Kivisild, T., Metspalu, E., Kaldma, K., Tambets, K., Tolk, H. V., Parik, J., Loogväli, E. L., Derenko, M., Malyarchuk, B., Bermisheva, M., Zhadanov, S., Pennarun, E., Gubina, M., Golubenko, M., Damba, L., Fedorova, S., Gusar, V., Grechanina, E., Mikerezi, I., Moisan, J. P., Chaventré, A., Khusnutdinova, E., Osipova, L., Stepanov, V., Voevoda, M., Achilli, A., Rengo, C., Rickards, O., De Stefano, G. F., Papiha, S., Beckman, L., Janicijevic, B., Rudan, P., Anagnou, N., Michalodimitrakis, E., Koziel, S., Usanga, E., Geberhiwot, T., Herrnstadt, C.,

Howell, N., Torroni, A., & Villems, R. (2003). Origin and diffusion of mtDNA haplogroup X. *American Journal of Human Genetics, 73*, 1178–1190.

Ronquist, F., Teslenko, M., van der Mark, P., Ayres, D. L., Darling, A., Höhna, S., Larget, B., Liu, L., Suchard, M. A., & Huelsenbeck, J. P. (2012). MrBayes 3.2: Efficient Bayesian phylogenetic inference and model choice across a large model space. *Systematic Biology, 61*, 539–542.

Roos, T., & Heikkilä, T. (2009). Evaluating methods for computer-assisted stemmatology using artificial benchmark data sets. *Literary and Linguistic Computing, 24*, 417–433.

Ross, R. M., Greenhill, S. J., & Atkinson, Q. D. (2013). Population structure and cultural geography of a folktale in Europe. *Proceedings of the Royal Society B, 280*, 1471–1529.

Sperber, D. (1996). *Explaining culture*. Oxford: Blackwell.

Stubbersfield, J. M., & Tehrani, J. (2013). Expect the unexpected? Testing for minimally counterintuitive (MCI) bias in the transmission of contemporary legends: A computational phylogenetic approach. *Social Science Computer Review, 31*(1), 90–102.

Stubbersfield, J. M., Tehrani, J. J., & Flynn, E. G. (2015). Serial killers, spiders and cybersex social and survival information bias in the transmission of urban legends. *British Journal of Psychology, 106*(2), 288–307.

Swofford, D. L. (1998). *PAUP* 4. Phylogenetic analysis using parsimony (*and other methods)*, Version 4. Sunderland: Sunderland.

Tehrani, J. (2013). The phylogeny of little red riding hood. *PLoS One, 8*(11), e78871.

Tehrani, J., & Collard, M. (2002). Investigating cultural evolution through biological phylogenetic analyses of Turkmen textiles. *Journal of Anthropological Archaeology, 21*, 443–463.

Tehrani, J., & Collard, M. (2009). On the relationship between inter-individual cultural transmission and population-level cultural diversity: A case study of weaving in Iranian tribal populations. *Evolution and Human Behavior, 30*, 286–300.

Tehrani, J., Nguyen, Q., & Roos, T. (2015). Oral fairy tale or literary fairy tale? Investigating the origins of little red riding hood using phylogenetic network analysis. *Digital Scholarship in the Humanities*. doi: 10.1093/llc/fqv016

Temkin, I., & Eldredge, N. (2007). Phylogenetics and material culture evolution. *Current Anthropology, 48*, 146–153.

Thompson, S. (1977). *The folktale*. California: University of California Press.

Toelken, J. B. (1969). A descriptive nomenclature for the study of folklore. Part I: The process of tradition. *Western Folklore, 28*, 91–101.

Uther, H. J. (2004). *The types of international folktales. Part I: Animal tales, tales of magic, religious tales and realistic tales, with an introduction*. Helsinki: Academia Scientiarum Fennica.

von Sydow, C. W. (1932). *Om traditionsspridning* (pp. 322–344). Scandia: Tidskrift för historik forskning.

von Sydow, C. W. (1948). Folktale studies and philology: Some points of view. In A. Dundes (Ed.), *The study of folklore* (pp. 219–242). Englewood Cliffs, NJ: Prentice Hall.

Ziolkowski, J. M. (1992). A fairy tale from before fairy tales: Egbert of Liege's "De puella a lupellis seruata" and the medieval background of "Little Red Riding Hood". *Speculum, 67*, 549–575.

Zipes, J. (1993). *The trials and tribulations of little red riding hood*. New York: Routledge.

Analyses of a Virtual World

Yurij Holovatch, Olesya Mryglod, Michael Szell, and Stefan Thurner

Abstract We present an overview of a series of results obtained from the analysis of human behavior in a virtual environment. We focus on the massive multiplayer online game (MMOG) *Pardus* which has a worldwide participant base of more than 400,000 registered players. We provide evidence for striking statistical similarities between social structures and human-action dynamics in real and virtual worlds. In this sense MMOGs provide an extraordinary way for accurate and falsifiable studies of social phenomena. We further discuss possibilities to apply methods and concepts developed in the course of these studies to analyse oral and written narratives.

Introduction

Quantitative approaches in social sciences and humanities have benefited greatly from the introduction of advanced information technologies. These allow one to accumulate and store a huge amount of data, as well as to enable its effective processing. Computer-based communication technologies have led to the formation of virtual societies, and these societies have themselves become the subjects of research. In this chapter, we demonstrate some results obtained through analyses of human behavior in a Massive multiplayer online game (MMOG) (Castronova 2005). Playing such games has become one of the largest collective human activities

Y. Holovatch • O. Mryglod (✉)
Institute for Condensed Matter Physics, National Academy of Sciences of Ukraine,
79011 Lviv, Ukraine
e-mail: hol@icmp.lviv.ua; olesya@icmp.lviv.ua

M. Szell
Center for Complex Network Research, Northeastern University, 02115 Boston, MA, USA
e-mail: m.szell@neu.edu

S. Thurner
Section for Science of Complex Systems, Medical University of Vienna, Vienna, Austria

Santa Fe Institute, Santa Fe, NM 87501, USA

IIASA, Schlossplatz 1, A-2361 Laxenburg, Austria
e-mail: stefan.thurner@meduniwien.ac.at

© Springer International Publishing Switzerland 2017
R. Kenna et al. (eds.), *Maths Meets Myths: Quantitative Approaches
to Ancient Narratives*, Understanding Complex Systems,
DOI 10.1007/978-3-319-39445-9_7

in the world; at present hundreds of millions of people participate in such activities including, for example, approximately 10 million who are registered for the most popular MMOG *World of Warcraft* (Statista 2015). In turn, the records of activity of players in MMOGs provide extraordinary opportunities for quantitative analyses of social phenomena with levels of accuracy that approach those of the natural sciences. The results we discuss below were obtained from a series of analyses of the MMOG *Pardus*. Since it was launched in 2004, the *Pardus* game served as a unique testing ground to measure different observables that characterize inhabitants of the virtual world and in this way to obtain clues also on complex social processes taking place in the real[1] world (Szell and Thurner 2010; Szell et al. 2010, 2012; Thurner et al. 2012; Szell and Thurner 2013; Klimek and Thurner 2013; Corominas-Murtra et al. 2014; Fuchs and Thurner 2014; Fuchs et al. 2014; Sinatra and Szell 2014).

The reasons for the appearance of a chapter on a multiplayer online world in a book devoted to complexity-science approaches to oral and written narratives may not be obvious at first sight. Comparative mythology, folktales and epic literature which are the main issues in this book have little to do with the virtual world of *Pardus*. However, a more careful comparison reveals a number of common features and potentially transferrable analytical tools. In both cases, one treats narrative or virtual characters in a manner similar to how sociologists treat real social groups, with an aim to quantify properties of such groups. See e.g. Stiller et al. (2003), Mac Carron and Kenna (2012); MacCarron and Kenna (2013) and references therein. In such studies, quantitative analyses put comparison and classification of different narratives on a solid basis. A similar goal is pursued by the analysis of actions of virtual characters (players) of an MMOG.

Although the societies of an MMOG and of a narrative are to some extent mirrors of the real world, they reflect it in different ways. In an MMOG, each individual is the character controlled by a player (i.e., an avatar or graphical representation of the user). In a narrative, the individual is a character in a story. The narrative is created with the intention to be perceived by a reader and it carries a personal contribution of a writer. Life in an MMOG evolves as a complex system and is driven by numerous interactions between players. Here we discuss some results of analyses of the virtual world and methods used to obtain them which, we hope, might be also useful in future analyses of the world of narratives. Moreover, analysis of life in a synthetic world serves as a tool to learn more about human behavior in the real world.

We discuss the application of complex-network concepts to uncover the diversity of social interactions in a society. We pay particular attention to how multidimensional graphs, wherein nodes may be connected by more than one type of edge or link, contribute to the formation of different interconnected *multiplex* networks. We demonstrate how one can test traditional social-dynamical hypotheses which apply to virtual societies too, bringing to the fore intrinsic similarities between virtual and real worlds. In addition, we analyse the evolution of social networks in time through

---

[1]We use the word "real" due to lack of a better term. Certainly human behavior, emotions, and decisions in online worlds are as "real" as in the offline world—they might only be biased differently depending on the context.

a first analysis of dynamical features of multi-level human activity (sequences of human actions of different types). This study of multi-level human activity in section "Human Multi-Level Activity" can be seen as a dynamic counterpart to static multiplex network analysis presented in the following section.

Database and Networks

Pardus is a browser-based MMOG played since September 2004. It is an open-ended game with no explicit winning purpose and a worldwide player base of more than 400,000 registered participants (Pardus 2015; Szell and Thurner 2010). The game has a science fiction setting and each player controls one character. The characters act within a virtual world, making up their own goals and interacting with the social environment which is self-organized. The game features three different universes: *Orion, Artemis,* and *Pegasus.* Each universe has a fixed start date but no scheduled end date. The results we discuss here concern the *Artemis universe,* selected for study because it has most active players and because its data set is most complete. Artemis was opened on June 10, 2007 and at the time of this study was inhabited by several thousand active characters.

Each character in the game is a pilot who owns a spacecraft, travels in the universe and is able to perform a number of activities of different types, such as communication, trade, attack, establishing or breaking friendships or enmities, etc. Since we focus on social features, we make use of records about the following activities of each character in the game:

- sending private messages from one player to another (communication, C);
- attacking other players or their belongings (attack, A);
- trading or giving gifts (trade, T);
- indicating friends by adding their names to a friend list (F);
- indicating enemies by adding their names to an enemy list (E);
- removing friends from the friend list (D);
- removing enemies from the enemy list (X);
- placing a bounty on other players (B).

The overall number of actions performed by the characters during 1238 consecutive days of observation was $N = 8,373,209$ [for a detailed description of the database see Szell and Thurner (2010), Mryglod et al. (2015)].

A straightforward way of mapping the *Pardus* society onto a complex network is to associate nodes with individual characters. A link between two nodes represents an action that took place between the corresponding pair of characters. Every action type (from the above list) is directed; it is initiated by one character and directed towards another. Given the different possible actions, one arrives at a set of directed networks where links in each network correspond to actions of certain types. We define the in- and out-degrees as the number of incoming and outgoing links that a given node has, and we denote these by k^{in} and k^{out} respectively. When we don't

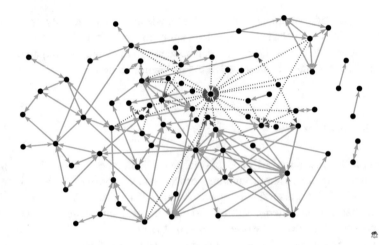

Fig. 1 Friendship (*green, solid*) and enemy (*red, dashed*) relations on day 445 between 78 randomly selected game characters. The diagram is taken from Szell and Thurner (2010). An animated evolution of this network can be seen at: http://www.youtube.com/user/complexsystemsvienna

specify whether we are dealing with in- or out-degrees, or if it doesn't matter (when the network is not directed), we use the generic symbol k for the degree instead. Data are available with a one second resolution making possible a refined analysis of dynamical features of the virtual society. With the data on activities of each player to hand, one can construct networks of social interactions at each instant in time and follow their evolution.

As an example, in Fig. 1 we show networks of friendship and enemy relations on the 445th day (01 September 2008) between 78 randomly selected characters. One can measure the basic network properties, track their evolution over time (Szell and Thurner 2010) and quantify correlations between properties of networks of different types (Szell et al. 2010). In Fig. 2 we show several features of the communication (C), friendship (F), and enemy (E) networks, measured during the same day in the virtual world. We display three measures: the cumulative degree distribution $P(k)$, the clustering coefficient $C(k)$ and the mean degree $k^{nn}(k)$. The first of these, $P(k)$, is the probability that the degree k_i of a randomly selected node i is at least as large as a given value k ($k_i \geq k$). The second measure, $C(k)$, is defined in the following manner. One first forms the ratio of the number of links which *actually* exist between i's neighbours and the number of all *possible* links between them. If node i has k_i neighbours, each of these can be linked to $k_i - 1$ other neighbours of node i. The total number of potential links between i's neighbours is therefore $k_i(k_i - 1)/2$, having divided by two to deal with the overcounting induced by each link being shared by two nodes. If y_i is the actual number of links between the neighbours of node i, then we define the clustering of the ith node to be

$$C_i = \frac{2y_i}{k_i(k_i - 1)}, k_i > 1. \tag{1}$$

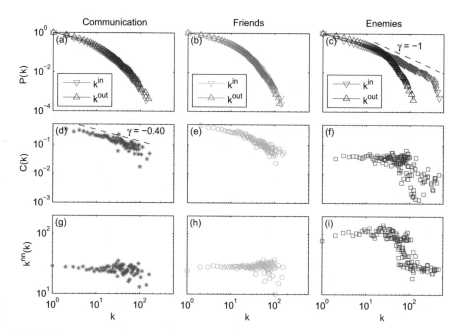

Fig. 2 Cumulative degree distributions of (**a**) communication (C), (**b**) friendship (F) and (**c**) enemy (E) networks; clustering coefficient C as a function of degree k for the (**d**) C, (**e**) F and (**f**) E networks; nearest neighbor mean degree k^{nn} versus degree of the (**g**) C, (**h**) F, and (**i**) E networks. Fits to power laws ($\sim k^\gamma$) are indicated by *dashed lines*, when appropriate. All distributions are shown as for 2008-09-01, the picture is taken from Szell and Thurner (2010)

Taking the average of the C_i's over all nodes for which $k_i = k$ gives the mean clustering $C(k)$. This is the average clustering associated with nodes of degree k. The third and final measure in Fig. 2, $k^{nn}(k)$, is the mean degree of the nearest neighbours of nodes which themselves have degree k. These and other basic network properties of the *Pardus* society are discussed in detail in Szell and Thurner (2010) and further analyzed in numerous publications (Szell et al. 2010, 2012; Thurner et al. 2012; Szell and Thurner 2013; Klimek and Thurner 2013; Corominas-Murtra et al. 2014; Fuchs and Thurner 2014; Fuchs et al. 2014; Sinatra and Szell 2014; Mryglod et al. 2015).

Figure 2 tells us that there are a number of characteristics that distinguish networks of different types. Each plot is on a double-logarithmic scale so that any power-laws present would show up as straight-line segments. For example, comparing the properties of the communication, friendship, and enemy networks, one finds that only in the latter case can the cumulative node degree distribution be approximated by a power law. This is evidenced in Fig. 2c where the fit indicates that $P(k) \sim k^\gamma$ with $\gamma = -1$. The corresponding plots for the communication and friendship cases, Fig. 2a, b, respectively, are evidently not described by power-laws.

Figure 2d, e shows that the clustering coefficients $C(k)$ for the communication and friendship networks exhibit clear downward trends as k increases, whereas

Fig. 2f shows that the clustering coefficients are to a large extent independent of k for the enemy networks, at least for large values of k. That the same can be said about the degrees k^{nn} is evident from Fig. 2g–i.

These observations can be complemented by examining the behavior of the *linking probability* $p(k)$ as function of the degree k (not shown in Fig. 2). This is the probability of a new node connecting to an existing node with degree k. By fitting it with a power-law function of the type $p(k) \sim k^{\alpha}$, we can try to understand how the network grows as new nodes are added. If α is positive, it means that new nodes prefer to attach to nodes which have higher degrees. In the case where $\alpha = 1$, which signals a linear dependency, this phenomenon is known as *preferential attachment*. The following values for the exponents have been reported for in-degrees: $\alpha \simeq 0.62$ for the friendship and $\alpha \simeq 0.90$ for the enemy network. We refer to Szell et al. 2010 for further discussions on this topic.

These quantitative observations are important because they allow one to conclude that there are intrinsic differences in the network formation processes for the C, F and E networks. In particular, in a typical preferential attachment mechanism, links attach to nodes according to how many links these nodes already have; high-degree nodes receive more new neighbours than their low-degree counterparts. The resulting node degree distribution decays as a power law, the linking probability increases linearly with k (i.e. $\alpha = 1$) and the clustering coefficient $C(k)$ is uniform as a function of k (Barabási and Albert 1999). None of these three properties holds for the communication or friendship networks, but all are roughly satisfied for the enemy network. This suggests that the preferential attachment scenario is mainly relevant for the latter. In other words, the more enemies a character has, the more likely they are to accrue more enemies but the same is not true for friends or for communication.

The above discussion around Fig. 2 focuses on the differences between the various network types. There are similarities too. The following common features have been observed for *Pardus* networks of different types (Szell and Thurner 2010): (i) their average node degrees grow over time; (ii) the mean shortest path length decreases with time; (iii) networks such as communication, trade and friendship, are reciprocal (individuals tend to reciprocate connections); (iv) but networks such as enmity, attack and bounty are not. Properties (i) and (ii) signal that the network becomes more dense as time evolves while properties (iii) and (iv) show that networks with positive connotations tend to be reciprocal while those with negative connotations are not.

The example network shown in Fig. 1 is a part of a multiplex network (Wasserman and Faust 1994), that consists of a set of characters that are joined by different types of links, corresponding to different types of social relations (recorded as different actions, in the case of the *Pardus* game). It is well established by now that multiplexity plays an essential role in network organization. Indeed, an interplay between different social relations, expressed as an interaction of links of different types, may lead to new levels of complexity. To quantify the relations between multiplex network layers, a thorough analysis for the *Pardus* society has focused

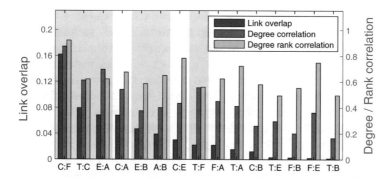

Fig. 3 Link overlap, degree correlation, and degree rank correlation for different pairs of networks. Pairs of equal connotation (positive-positive or negative-negative) are marked with a *gray* background. See Szell et al. (2010) for more explanations, where from the picture has been taken

on the *link overlap* and correlations between node degrees between the different network layers (Szell et al. 2010).

The *link overlap* quantifies the interaction between two networks by measuring the tendency that edges are simultaneously present in both networks. It is defined by the so-called *Jaccard coefficient*, which is a similarity score between two sets of elements. It is defined as the size of the intersection of the sets divided by the size of their union. It is therefore a global measure which ranges in size from zero (no overlap) to 1 (100 % overlap).

The correlation between node degrees (or their ranks), on the other hand, measures the extent to which degrees of agents in one type of network correlate with degrees of the same agents in the other one. If the correlation between node degrees in two different networks is close to 1, players who have many links in one network have many links in the other one and vice versa. In Fig. 3 we show these quantities for different pairs of networks. One sees that pairs of the same connotation (positive-positive or negative-negative) typically have high overlaps, whereas this is not the case for pairs of opposite connotation. Moreover, low values of the degree correlations indicate that hubs in one network are not necessarily hubs in another. This demonstrates the tendency of individuals to play different roles in different networks.

Properties of the multiplex network vary for different types of players. A recent analysis of gender-specific differences has shown (Szell and Thurner 2013) that females and males manage their social networks in substantially different ways. In particular, on the individual level, females perform better economically and are less inclined to take risks than males. Males reciprocate friendship requests from females faster than vice versa and hesitate to reciprocate hostile actions of females. On the network level, females have more communication partners, who are themselves less connectcd than partners of males. Cooperative links between males are under-represented, reflecting competition for resources among males.

Analysis of the *Pardus* universe also allows one to quantify to what extent classical sociological hypotheses hold up in a virtual world. Such analyses recently enabled us to propose two approximate social laws in communication networks (Szell and Thurner 2010). These findings were made in the course of testing Granovetter's *Weak Ties Hypothesis* (Granovetter 1973), which suggests casual acquaintanceships link communities in an essential way. This means that weak links are important to hold the network together—it is the weak links, not the strong ones, that tend to form the ties between distinct sets of nodes. To explain this quantitatively, we require three new concepts. Firstly, the *overlap O* is the fraction of common neighbours between two neighbouring nodes; if A and B represent the sets of neighbours of two nodes, the overlap is the number of nodes in that A and B have in common $(A \cap B)$ divided by the total number in A and B taken together $(A \cup B)$. This is a local measure, distinct from the global link overlap discussed earlier. Secondly, the *link-betweenness centrality, b*, is the ratio of the number of shortest paths between two nodes that contain a given link to the total number of shortest paths between these nodes. Thirdly we need the *weight w* of a link joining two nodes. This is also a local quantity. For the communication network, the weight of a link between two nodes corresponds to the number of private messages sent between two individuals these nodes represent.

The stronger the connection between two individuals, the more similar is their local environment, and vice versa. Therefore we expect that the overlap is an increasing function of weight. Analysis of the structure of communication networks in the *Pardus* world has indeed revealed that, on average, the overlap O related to the weight w of a link joining two given nodes increases as

$$O(w) \sim \sqrt[3]{w}. \tag{2}$$

To understand how this connects with Granovetter's hypothesis, consider two sets of nodes: one set involving Node A and one including Node B. If the set of nodes connected to Node A is very distinct from the set of nodes connected to Node B, then the overlap O between the two sets is low (it is zero if they are completely distinct). If O is small, Eq. (2) tells us that w is also a small number. This means that the weight w between nodes A and B is small on average. This low weight corresponds to the notion of casual relationship. Thus Eq. (2) quantifies Granovetter's hypothesis—it tells us that light-weight relationships are essential to bind distinct sets of nodes.

Another way to quantify the Weak Ties Hypothesis is to check the behavior of the overlap O as a function of link-betweenness centrality, b. If the hypothesis is valid, shortest connections between two sets of nodes are forced to go through the weak links that connect them. In other words, low overlap corresponds to high betweenness. The obtained dependency was indeed found to be a decreasing function of the explicit form

$$O(b) \sim \frac{1}{\sqrt{b}}, \tag{3}$$

which supports the hypothesis.

Other social hypotheses that were tested and confirmed for the *Pardus* networks concern triadic closure and network densification (Granovetter 1973). The triadic closure conjecture follows balance considerations (Heider 1946) and reflects the property among three nodes A, B, and C in a social network, that if node pairs A-B and A-C are linked by strong ties, there tends to be a weak or strong tie between the node pair B-C. The phenomenon of triadic closure (Rapoport 1953) states that individuals are driven to reduce the cognitive dissonance caused by the absence of a link in the (unclosed) triad. Because of this the triad in which there exist strong ties between all three subjects A, B and C should appear in a higher than expected frequency. Network densification (i.e. shrinking of its diameter and growing of average degrees with a span of time) is an aging effect that has been observed recently in many growing networks (Leskovec et al. 2007). Observation of similar effects for *Pardus* networks serves as one more argument about universal features of this phenomenon. It is worth noting one feature known in the real world that is also reflected in the virtual society. It concerns the number of people with whom one can maintain stable social relationships, given humans' limited cognitive capacities. This is the so-called "Dunbar" number (Dunbar 1993). See also Kenna and Berche (2010) for a mathematical basis for the upper limit of group sizes. A prominent feature of the plots in Fig. 2 is that the maximal out-degree of networks represented there is limited by $k^{out} \simeq 150$, a value conjectured to be the maximal number of stable relationships humans can comfortably maintain (Dunbar 1993).[2]

Results discussed so far give a quantitative description of the *Pardus* society based on a network perspective. We argued how networks of different social interactions arise and evolve, how they interact with one another, what are the observables that describe their properties and what are their implications for life in a virtual world. Although dynamical features of network evolution were also analyzed here, we did not address so far the question of temporal structure of human actions. We now ask if there exist regularities that govern temporal behavior of characters in a virtual world and if so, do they resemble those in the real world? Some answers to this question will be given in the following section.

Human Multi-Level Activity

The lives of humans can be viewed as sequences of different actions. Some of these are performed on a regular basis; others have strong stochastic components. Some actions are performed frequently; others are carried out sporadically. One associated quantity of interest is the time-lapses between such events—the "inter-event" times. Many early models which were used to study the inter-event time distribution of human action sequences were based on the assumption that such actions are performed randomly in time. In the simplest cases, these are described statistically

---

[2]Editors'note: See Robin Dunbar's chapter in this volume.

Player 8   ... CC C CT CCBCCCCCCCC   CCCCC C C CC ...
Player 676 ... FFFF CCCCC   CFFCCCCC   CC FCTCCCEE ...
Player 784 ... AAAAAAAAAA   T T T T T   CCC CC ...
Player 910 ... CFC C C C TT   TTT T F T CC   C   CT   E T   ...

Fig. 4 Short segment of action sequences, performed by four players. Different actions are shown by *different letters* as explained in section "Database and Networks"

by a Poisson process. This assumption suggests that times between actions of the same individual are independent and distributed exponentially. Models of such kind are still being used, however there appears to be an accumulation of evidence that distribution functions characterizing sequences of different human actions in time are highly non-trivial [see Barabási (2005), Oliveira and Barabási (2005), Vazquez et al. (2006), Malmgren et al. (2009), Goh and Barabási (2008), Wu et al. (2010), Jo et al. (2012), Yasseri et al. (2012a,b) and references therein]. An inhomogeneous bursty distribution of human actions influences their temporal statistics and often is associated with power laws. Such conclusions were made on the basis of observing different types of single human action such as writing letters, checking out books in libraries, writing e-mails, web browsing, and many more. Analysis of temporal features of the performance of actions of different types, which we call a multi-level human activity, still remains an open challenge. The main problem here is the obvious difficulty in accessing reliable and statistically relevant databases of records of various forms of human activity.

The clear advantage of our data set on the multi-level activity of characters in the *Pardus* world is that it is based on the analysis of behavior of thousands of characters across several years, and that it concerns various types of actions. In this sense it can be considered as the dynamic counterpart of static multiplex network analysis. The main outcome of this study is given in this section. The interested reader is referred to Mryglod et al. (2015) for a more extensive report.

Figure 4 shows four segments of action sequences, performed by four *Pardus* players. Different actions are shown by different letters as explained at the beginning of section "Database and Networks". The times for each action have been recorded which allows us to analyze peculiarities of temporal behavior of each player during the whole observation time (for the results shown below it is equal to 1238 days since 10 June 2007 when the *Artemis universe* was opened). Further, we can assemble a general picture of temporal behavior of all players. Below we concentrate on the statistics of inter-event times τ, i.e. the time intervals between two consecutive actions of the same player. In Fig. 5 we show the distribution functions of the inter-event time τ for all actions[3] of all players who performed at least 50 actions, considering players with fewer actions being not representative. As it can be

---

[3]Here we take into account all actions as listed at the beginning of section "Database and Networks", discarding for technical reasons the 'bounties' B.

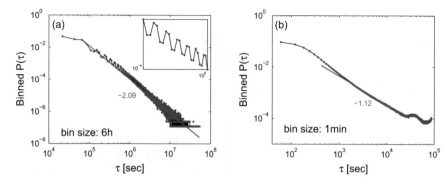

Fig. 5 Distribution of the inter-event times τ for all players who performed at least 50 actions. (**a**) Entire observation period (1238 days), bin size is 6 h=21,600 s. (**b**) First 24 h, bin size is 1 min. *Inset*: same as (**a**) for 6 days. Circadian rhythms are clearly visible. The picture is taken from Mryglod et al. (2015)

seen from Fig. 5a, b, the distributions follow approximate power laws, the numerical value of the exponent depending on the chosen bin size. A prominent feature of the plots is that they manifest a fine structure: one observes regular patterns of various periodicity when the plots are considered on a smaller scale. The emergence of periodic patterns is known to be an inherent feature of human activity (Jo et al. 2012; Yasseri et al. 2012a; Malmgren et al. 2009). In our case it can be naturally explained by circadian and active working day cycles as well as by peculiarities of performing different actions (Mryglod et al. 2015).

The power-law behavior of inter-event time distribution functions signals the bursty nature of human dynamics (Barabási 2005; Oliveira and Barabási 2005). One of the variables that is used to quantify burstiness is the so-called burstiness index (Goh and Barabási 2008), B, defined as (Jo et al. 2012; Yasseri et al. 2012b)

$$B \equiv \frac{\sigma - m}{\sigma + m}, \tag{4}$$

where m and σ represent the average inter-event time and the standard deviation, respectively. As follows from definition in Eq. (4), a value of $B \approx -1$ characterizes regular patterns. Random Poisson process with a fixed event rate yields $B \approx 0$. In Fig. 6 we show several action streams of individual players with different values of burstiness. Panels a, b, and c display action streams where B is maximal, minimal, and close to zero. We find that the average value of burstiness for all actions of all players is $\overline{B} \simeq 0.53$. This feature of the virtual world can be compared with the burstiness values characterizing real world activities of mobile communication $\overline{B} \simeq 0.2$ (Jo et al. 2012), and Wikipedia editing[4] $\overline{B} \sim 0.6$ (Yasseri et al. 2012b). Note that, similar to the real world, the burstiness is action-specific in the virtual world

---

[4]In this case, the events correspond to consequent edits of Wikipedia articles.

Fig. 6 Action streams of
three players, each with
different values of burstiness
B. *Lines* mark times of
executed actions, the *distance
between lines* is the
inter-event time

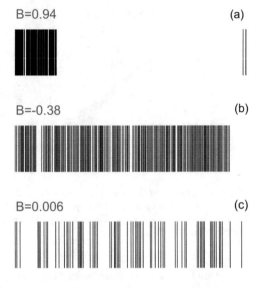

too. We find that burstiness of attacks has larger values than for communication. This illustrates the intuitive understanding of the nature of these actions: attacks (A) appear highly clustered within short time intervals, while communication (C) is more uniformly distributed over time.

Another inherent feature of the multi-level human activity in the *Pardus* world is that time distributions found there are action-specific: each type of actions has a particular and characteristic distribution. This fact is far from trivial, the claim being that the decay of the inter-event time distributions might serve as a distinguishing feature of action type. Similar to physics, where one can classify life times of unstable elements by decay constants that uniquely characterize each element, one can quantify the decay of action-specific inter-event time distributions by the decay constants that uniquely characterize different types of actions too. Leaving the detailed discussion of this fact to a separate publication (Mryglod et al. 2015) where the inverse cumulative distributions $P(\geq \tau)$ of inter-event times were analyzed and the numerical values of the decay constants can be found, we mention here, that the overall behavior of players is characterized by three different time scales: (i) an immediate reaction (τ does not exceed several minutes), (ii) an early day (τ is less then 8 h), (iii) a late day (τ is between 8 and 24 h). At long times (more then several months) an exponential cut off becomes apparent, whereas for very short times (time scale (i)) all distributions have a similar tendency to decay very fast: the short inter-event times are typical for most of the actions. The scales (ii) and (iii) bring about specific features of different actions and allow to classify them. We found that on scale (ii) the inverse cumulative distributions of inter-event times of every specific action are best approximated by a power law:

$$P(\geq \tau) \sim \tau^{-\alpha}, \tag{5}$$

Fig. 7 Number of actions of all types per day over time. There are four pronounced peaks in the player activity that corresponds to specific events that happened in the virtual world during the observation period: the *three coloured vertical stripes* indicate war periods, the *thin vertical line* indicates the introduction of a major new game feature

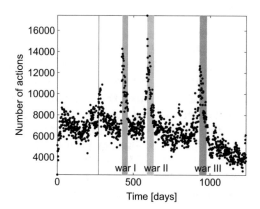

while on scale (iii) the decay is of exponential form:

$$P(\geq \tau) \sim \exp\left(-\tau/\tau_0\right). \tag{6}$$

Constants α and τ_0 allow to discriminate between actions of different type and to quantify them in the unique way.

Before finishing this chapter we comment on the global dynamics and activity patterns that can be observed in the *Pardus* world. In Fig. 7 we show how actions are distributed over time. One can see four pronounced peaks in the players' activities. They correspond to specific events that happened in the virtual world during the observation period: the three coloured vertical stripes in the figure indicate war periods, the vertical line indicates the introduction of a major new game feature. Besides an obvious conclusion about the increase of activity during war periods, changes in action-specific dynamics are observed (e.g., intensification of attacks and communication). Another question of interest was to check whether the changes in player activity might serve as precursors of coming wars (or precursors of the end of war). Interaction and coexistence of different social relations are important to describe conflicts in social systems (Bohorquez et al. 2009; Lim et al. 2007; Clauset and Gleditsch 2012). Assuming that a war in a virtual world emerges and finishes as a result of a complex process of social interaction, it is tempting to ask about details of early and late stages of this process. Our attempts to use cross-correlation analysis for finding potential lead effects of player activity patterns on the onset of war have not (yet) lead to conclusive answers (Mryglod et al. 2015).

Conclusions and Outlook

To what extent is the behavior of characters in a virtual world similar to human behavior in the real world? To answer this question, one has to compare quantitative behavioral features in both worlds. Results obtained in the analysis of player behavior in the MMOG *Pardus* provide solid evidence for the existence of certain similarities of social structure and human dynamics in the real and the virtual worlds. These similarities concern certain types of social networks, their growth patterns, the validity of major sociological hypotheses, gender specific dynamics, etc. In this sense, MMOGs provide an extraordinary opportunity for an accurate analyses of social phenomena and falsifiable hypotheses.

Let us return to the question about the comparison of social activities in a virtual world and in the (written or oral) narrative. Two properties of the virtual society are obvious: (i) its structure changes over time, (ii) every element of the society (every character) acts in time. To what extent might the study of these properties be useful for similar analyses of narratives? In section "Database and Networks" we have shown how property (i) is covered within the network formalism. In addition, one may use inter-event time distributions to quantify property (ii). To this end, the application of a multi-level human activity formalism may be useful, as outlined in section "Human Multi-Level Activity". In the analysis of social networks of narratives, usually the resulting static networks of all acting characters are studied (Mac Carron and Kenna 2012). In principle, one can get access to their evolution too, introducing the time-line via counts of narrative subunits [pages or chapters (MacCarron and Kenna 2013), or appearance of new actors (Dunbar 1993)]. In this sense, property (i) is accessed in the narrative analysis too. We are not aware of analyses of property (ii) for a narrative. Although such a problem statement might be interesting, its realisation, besides obvious difficulties in introducing a coherent and self-consistent time line, will meet difficulties of separating dynamics caused by the evolution of a subject and a style of presentation.

Acknowledgements We want to express our thanks to the Editors of the book, Ralph Kenna, Máirín MacCarron, and Pádraig MacCarron, for the invitation to write this chapter and for useful suggestions and to Anita Wanjek for helpful comments on the manuscript. This work was supported in part by the 7th FP, IRSES project No. 612707 Dynamics of and in Complex Systems (DIONICOS) and by the COST Action TD1210 Analyzing the dynamics of information and knowledge landscapes (KNOWSCAPE). ST acknowledges support by the EU FP7 project LASAGNE no. 318132.

References

Barabási, A. (2005). The origin of bursts and heavy tails in human dynamics. *Nature, 435*, 207.
Barabási, A., & Albert, R. (1999). Emergence of scaling in random networks. *Science, 286*, 509.
Bohorquez, J. C., Gourley, S., Dixon, A. R., Spagat, M., & Johnson, N. F. (2009). Common ecology quantifies human insurgency. *Nature, 462*, 911.
Castronova, E. (2005). *Synthetic worlds. The business and culture of online games* (332 pp.). Chicago: The University of Chicago Press.
Clauset, A., & Gleditsch, K. S. (2012). The developmental dynamics of terrorist organizations. *PLoS ONE, 7*(11), e48633.
Corominas-Murtra, B., Fuchs, B., & Thurner, S. (2014). Detection of the elite structure in a virtual multiplex social system by means of a generalised K-core. *PLoS ONE, 9*(12), e112606.

Dunbar, R. (1993). Coevolution of neocortical size, group size and language in humans. *Behavioral and Brain Sciences, 16*(4), 681.

Fuchs, B., Sornette, D., & Thurner, S. (2014). Fractal multi-level organisation of human groups in a virtual world. *Scientific Reports, 4,* Art 6526.

Fuchs, B., & Thurner, S. (2014). Behavioral and network origins of wealth inequality: Insights from a virtual world. *PLoS ONE, 9*(8), e103503.

Goh, K., & Barabási, A. L. (2008). Burstiness and memory in complex systems. *Europhysics Letters, 81,* 48002.

Granovetter, M. (1973). The strength of weak ties. *American Journal of Sociology, 78*(6), 1360.

Heider, F. (1946). Attitudes and cognitive organization. *Journal of Psychology, 21*(2), 107–112.

Jo, H.-H., Karsai, M., Kertész, J., & Kaski, K. (2012). Circadian pattern and burstiness in mobile phone communication. *New Journal of Physics, 14,* 013055.

Kenna, R., & Berche, B. (2010). The extensive nature of group quality. *Europhysics Letters, 90,* 58002.

Klimek, P., & Thurner, S. (2013). Triadic closure dynamics drives scaling laws in social multiplex networks. *New Journal of Physics, 15,* 063008.

Leskovec, J., Kleinberg, J., & Faloutsos, C. (2007). Graph evolution: Densification and shrinking diameters. *ACM Transactions on Knowledge Discovery from Data, 1*(1), 2.

Lim, M., Metzler, R., & Bar-Yam, Y. (2007). Global pattern formation and ethnic/cultural violence. *Science, 317,* 1540.

Mac Carron, P., & Kenna, R. (2012). Universal properties of mythological networks. *Europhysics Letters, 99,* 28002.

MacCarron, P., & Kenna, R. (2013). Network analysis of the Islendinga sogur - The Sagas of Icelanders. *European Physical Journal B, 86,* 407.

Malmgren, R. D., Stouffer, D. B., Campanharo, A. S. L. O., & Amaral, L. A. N. (2009). On universality in human correspondence activity. *Science, 325,* 1696.

Mryglod, O., Fuchs, B., Szell, M., Holovatch, Yu., & Thurner, S. (2015). Interevent time distributions of human multi-level activity in a virtual world. *Physica A, 419,* 681.

Oliveira, J. G., & Barabási, A. (2005). Darwin and Einstein correspondence patterns. *Nature, 437,* 1251.

Pardus. (2015). *Web-page of the Pardus game.* Retrieved May 14, 2015, http://www.pardus.at

Rapoport, A. (1953). Spread of information through a population with socio-structural bias. I. Assumption of transitivity. *Bulletin of Mathematical Biology, 15*(4), 523–533.

Sinatra, R., Szell, M. (2014). Entropy and the predictability of online life. *Entropy, 16,* 543.

Statista. (2015). Web-portal Statista (2015). *World of WarCraft subscribers by quarter.* Retrieved May 14, 2015, http://www.statista.com/statistics/276601/number-of-world-of-warcraft-subscribers-by-quarter/

Stiller, J., Nettle, D., & Dunbar, R. I. M. (2003). *Human nature* (Vol. 14, No. 4, pp. 397–408). New York: Walter de Gruyter, Inc.

Szell, M., Lambiotte, R., & Thurner, S. (2010). Multirelational organization of large-scale social networks in an online world. *Proceedings of the National Academy of Sciences of the United States of America, 107,* 13636.

Szell, M., Sinatra, R., Petri, G., Thurner, S., & Latora, V. (2012). Understanding mobility in a social petri dish. *Scientific Reports, 2,* 457.

Szell, M., & Thurner, S. (2010). Measuring social dynamics in a massive multiplayer online game. *Social Networks, 32,* 313.

Szell, M., & Thurner, S. (2013). How women organize social networks different from men. *Scientific Reports, 3,* 1214.

Thurner, S., Szell, M., & Sinatra, R. (2012). Emergence of good conduct, scaling and Zipf laws in human behavioral sequences in an online world. *PLoS ONE, 7*(1), e29796.

Vazquez, A., Oliveira, J. G., Dezso, Z., Goh, K. I., Kondor, I., & Barabasi, A. L. (2006). Modeling bursts and heavy tails in human dynamics. *Physical Review E, 73*(3), 036127.

Wasserman, S., & Faust, K. (1994). *Social network analysis: Methods and applications* (pp. 37–48). Cambridge: Cambridge University Press.

Wu, Y., Zhou, C., Xiao, J., Kurths, J., & Schellnhuber, H. J. (2010). Evidence for a bimodal distribution in human communication. *Proceedings of the National Academy of Sciences of the United States of America, 107*, 18803.

Yasseri, T., Sumi, R., & Kertész, J. (2012a). Circadian patterns of Wikipedia editorial activity: A demographic analysis. *PLoS ONE 7*(1), e30091.

Yasseri, T., Sumi, R., Rung, A., Kornai, A., & Kertész, J. (2012b). Dynamics of conflicts in Wikipedia. *PLoS ONE 7*(6), e38869.

GhostScope: Conceptual Mapping of Supernatural Phenomena in a Large Folklore Corpus

Peter M. Broadwell and Timothy R. Tangherlini

Abstract Since the inception of the field of folkloristics in the early nineteenth century, scholars have paid considerable attention to the relationship between place and folklore. An important yet largely overlooked question is how individuals in a tradition group, through their storytelling, negotiate the conceptualization of their local environment. In our work, we present a preliminary method for representing the conceptual mapping of the environment by storytellers or classes of storytellers. As opposed to our other geographically based methods for exploring the corpus, in which places are anchored in geography, *GhostScope* imagines all of the storytellers positioned at a conceptual "center." Geographic locations are then calculated for direction and distance based on this zero-point. The work is based on a digitized subset of legends from Evald Tang Kristensen's larger collection of Danish folklore.

Introduction: The "Witching Distance"

On a blustery afternoon in early March 1887, the folklore collector Evald Tang Kristensen (ETK) made his way down a small road in the Danish village of Hårby to a tiny dwelling that was the home of Ane Marie Jensdatter, an elderly woman who had been recommended to him by the local schoolteacher as an engaging

P.M. Broadwell
Digital Library, Charles E. Young Research Library, UCLA, Box 951575, Los Angeles, CA, USA

T.R. Tangherlini (✉)
The Scandinavian Section and the Department of Asian Languages and Cultures, UCLA, Box 951537, Los Angeles, CA, USA
e-mail: tango@humnet.ucla.edu

© Springer International Publishing Switzerland 2017
R. Kenna et al. (eds.), *Maths Meets Myths: Quantitative Approaches to Ancient Narratives*, Understanding Complex Systems,
DOI 10.1007/978-3-319-39445-9_8

storyteller.[1] Among the two dozen or so stories she told that afternoon was the following:

> De skyldte heksemesteren i Stjær for, at han malkede køerne i byen. Et sted tjente der en pige, som drömte, at hun skulde til at kjærne, og da kom denher heks, og så var det hele arbejde forgjæves. Hun siger så til konen om morgenen : "Skal vi til at kjærne i dag?" Konen siger ja. "Da har a drömt sådan og sådan", og hun fortalte nu det hele. Men det brød konen sig ikke om. Ligesom de skulde til at kjærne, kom manden. Så vilde konen have sat kjærnen af vejen, men hun turde ikke for hendes mand, og så fik de heller ikke smör den dag (DS VII 659).
>
> [The witch master from Stjær was blamed for milking the cows in the town. In one place, there was a hired girl who dreamed that, when she was going to churn, this witch came, and then all her work was for naught. She says to her mistress in the morning, "Are we going to churn today?" The woman says yes. "I dreamed so and so," and she told her the whole thing. The mistress wasn't too concerned about that. Just as they were about to churn, the witch master came along. She wanted to put the churn aside, but she didn't dare because of her husband, and then they didn't get butter that day either.]

The story recounts a noteworthy yet not uncommon encounter with the supernatural—here, a mildly threatening witch who makes his presence felt by disrupting the butter churning. Perhaps more important than the event itself is the close attention that Ane Marie pays to local places in her story.

By itself, Ane Marie's attention to place is not remarkable, since most of the stories ETK collected from smallholders, day laborers, and the rural poor in his decades of collecting include frequent mention of places (Tangherlini 2015). To most people, it may seem insignificant that the witch comes from Stjær, a village about four and a half kilometers to the northeast of Ane Marie's village of Hårby. But if one aggregates place name references such as those in Ane Marie's story with similar references from the thousands of other stories about witches that can be found in the collection—a task that can only be done accurately via computational methods—it turns out that she is not alone in ascribing a certain distance and direction to the threat of the witch.[2] By aggregating place name referents in stories, numerous other correlations between legend topics on the one hand and distance

---

[1] The target corpus for the current work is a digitized subset of approximately 35,000 legends from ETK's larger collection of Danish folklore. ETK, considered by many to be the single most prolific collector of folklore in world history, spent five decades traversing the Jutlandic peninsula, traveling largely on foot, collecting stories from the rural population and amassing a collection that spans 24,000 hand-written manuscript pages (Tangherlini 2013b). Unlike many of his contemporaries, ETK recorded metadata, including information about the storytellers and the communities where they lived. In all, he collected stories from approximately 3,500 named individuals. While the concentration of storytellers was greater in the countryside, he studiously avoided collecting in market towns and cities and, consequently, his storytellers are fairly evenly distributed throughout Jutland. For a discussion of the geographic distribution of storytellers and the normalization of story-related statistics to take into account differing population densities, see Broadwell and Tangherlini (2012).

[2] In a strange echo of Thor's exploits in Nordic mythology, where that god is often "out east" fighting not only giants but also shape shifting women (Hárbarðsljóð 37) and "brúðir bölvísar" [brides of malicious curses] (Hárbarðsljóð 23), there is a noticeable correlation in Danish legend tradition between witches and the east. In Hárbarðsljóð 37, Hlésey, the island home of the

and directionality on the other hand emerge. Consequently, one can begin to map the consensus of thousands of storytellers about the geographic locus of various elements of folk belief.

The tool we have devised, *GhostScope*, imagines that all of the storytellers are positioned at a conceptual "center." [3] Geographic locations are then calculated for direction and distance based on this zero-point. The user can choose classes of stories according to topic-driven indices; *GhostScope* then visualizes the aggregate relative distance and directionality of those topics based on the places mentioned in the stories and where the story was collected. It is intended to be part of a larger suite of tools in a *Folklore Macroscope* (Tangherlini 2013a) that includes corpus browsing tools such as the faceted browser, *ETKSpace*; a curated study environment such as the *Danish Folklore Nexus* (Tangherlini and Broadwell 2014), comprising the digital materials included with *Danish Folktales, Legends, and Other Stories* (Tangherlini 2013b); discovery tools such as *WitchHunter*, devised for geo-semantic mapping and pattern discovery (Tangherlini and Broadwell 2016); and *ElfYelp*, a tool for geo-topic modeling (Broadwell and Tangherlini 2015).

In the model of tradition that informs *GhostScope*, each story contributes in a very small but important way to the narrative encoding of the landscape (Gunnell 2008). Two major assumptions underlie this approach. First, storytellers comprise a close homogeneous group for which shared traditions act as a locus for the dynamic negotiation of the conceptualization of the lived environment. Over repeated tellings of stories, tradition dominants emerge, creating a shared conceptual map of belief, charting where threat and fortune lie in the landscape (Eskeröd 1947). Current theory recognizes that folklore emerges from the dialectic tension between individuals and tradition (Chesnutt 1999: 11). While individuals try to push the boundaries of what may be acceptable to the group, tradition is conservative, and exerts a cultural force that aims toward consensus. Through this process of give and take, legend, the genre on which we focus in this work, comes to act "as a symbolic representation of folk belief and collective experiences, and serves as a reaffirmation of commonly held values within the tradition group" (Tangherlini 2015: 40). Second, one must assume that the geographic referents on which storytellers rely for the localization and historicization of their accounts are not random, but play a significant role in creating this conceptual map (Tangherlini 2015: 11). In other words, one must assume that people telling and listening to these stories know where the places are to which they refer, and that they recognize, on some level, the association between places, events, and the beliefs that underlie those events.

Since the inception of the field of folkloristics in the early nineteenth century, scholars have paid considerable attention to the relationship between place and

---

shape-changing women (read witches) mentioned in Hárbarðsljóð 37, is generally identified as the northeastern Danish island of Læsø (Neckel 1983).

[3] *GhostScope*, along with the other macroscope tools described here, can be accessed at http:// etkspace.scandinavian.ucla.edu/macroscope.html

folklore (Grimm et al. 1816–1818). While the majority of the early attention focused on the nearly intractable question of the geographic origins of tales and motifs (Krohn 1926), more recent scholarship has focused on the relationship between storytellers and their physical environment (Gunnell 2008). An important yet largely overlooked question is how individuals in a tradition group, through their storytelling, negotiate the conceptualization of their local environment, coding certain areas as "safe", other areas as borderline or "liminoid", and yet other areas as "dangerous" (Turner 1983). These concepts of "inside", "outside" and "border" are fundamental to folk belief in general, and legend in particular, as are concepts such as "us" and "them" (Tangherlini 2015). In Denmark, the inside is closely related to concepts of "us" and is seen as "safe", while the outside is aligned with "them" (or the Other) and is seen as "dangerous" (Lindow and Tangherlini 1995). Border regions are areas of considerable ambiguity. We believe that meaningful clues as to how a particular cultural group conceptualized the environment in which they lived are encoded in these aggregate patterns.[4]

There are two key features of ETK's collections that allow *GhostScope* to provide meaningful views of these conceptual maps. First, storytellers frequently included externally identifiable place names in their stories, which has allowed us to geo-reference and disambiguate the majority of place names in the collection. These place names include both the places mentioned (PM) in the stories as well as the places where the stories were collected (places told or PT). Second, the collection is indexed in various ways, including by genre and topic. As noted, the genre of stories we have focused on are legends: short mono-episodic stories, told as true, that detail encounters with the supernatural or other forms of threat (Tangherlini 2015: 22). In his published collections, ETK classifies these stories into 36 main categories and 774 subcategories.[5] Although we primarily rely on the main categories in this work, the subcategories may be able to provide finer-grained representations of specific topics as we extend the tool.

Our method for representing the conceptual mapping of the environment by storytellers considers both relative direction and relative distance of encounters with

---

[4]Scholars of literary fiction also have explored the potential of plotting place references on maps to reveal intriguing aspects of the relationships between writers, readers, and their geographic environment (Moretti 1998). The advent of digitized literary corpora and software tools for text analysis and mapping have further enabled such inquiries, as well as related efforts to visualize fictional or conceptual geographies (Piatti et al. 2009). Large, digitized folklore collections offer researchers in the nascent field of "computational folkloristics" the opportunity to conduct even more in-depth analyses of the relationships between a folk tradition's participants and their highly specific temporal and physical settings (Tangherlini 2013a). In particular, it becomes possible not only to consider a "god's-eye view" of topic concentrations in the physical landscape, but also to place the storytellers at the center of a conceptual geography onto which their accounts of events in the surrounding environment can be superimposed, ultimately gaining a unique perspective on the storytellers' impressions of the world around them.

[5]In addition to this classification, we have devised an ontological classification scheme (Tangherlini 2013b) and a probabilistic classifier using Bayesian inference (Broadwell et al. 2015) although we have not included these alternative classifications in the work presented here.

threatening beings and other manifestations of folk belief (e.g. buried treasure). These representations of the stories' geographic information are coupled to the topic classification described above, thereby making it possible to consider how the "conceptual map" of the environment "looked" from the perspective of storytellers for any given topic. *GhostScope* allows users to pose questions such as, "Are there directions that are particularly associated with certain types of threat?" or, "Which supernatural beings keep their distance from the center and which come right up to the farm?" or, "Where are the conceptual borders in the landscape?"

Calibrating the GhostScope

As opposed to the other geographically based methods for exploring the corpus, which consider places as anchored in physical geography, *GhostScope* imagines all of the storytellers positioned at a conceptual "center" or zero-point. For each storyteller, this zero-point is their location when they told their stories (PT). Relative directions and distances to the places mentioned in the stories (PM) are then calculated based on this point. For example, in the first story presented above, Ane Marie lives in Hårby, which is set to zero, and the witch comes from Stæby, a point 4.5 km to the northwest. In the case of a different story, the storyteller lives in Henne, which is set to zero, and the witch he mentions destroys a farm in Sønder Bork, 13 km to the north. This process is repeated for each of the 20,431 stories in the target corpus that features a resolvable place name. Currently, the system only works with place names that can be geo-coded using available historical gazetteers. Consequently, place referents that are either toponyms known only to locals, and thus not included in gazetteers, or references to generic features of the natural environment (e.g. the hill behind Peder's house) or the man-made environment (e.g. Maren's barn) are not captured.[6]

The results of these aggregate calculations of places told to places mentioned (PT to PM) are represented through four main visualizations: (1) frequency graphs showing distance from the center in kilometers along the x-axis and number of stories along the y-axis, using both histograms binned[7] at 5 km intervals, and a line graph with 1 km binning (the overlapping Panels (a) and (b) in the top left of Fig. 1); (2) directional histograms or "wind rose" plots showing the frequencies of stories based on the relative direction of places mentioned in the story to the place where the stories were told, with 1° and 30° binning (Panels (c) and (d) in the top-right and middle-left panels of Fig. 1, respectively); (3) directional point-distribution maps that represent the number of stories mentioning places at a given distance and direction through the color and size of points on a 360° plot (Panel

---

[6]We reserve these challenges for future work.

[7]Statistical data binning involves the grouping together of continuous values into intervals called "bins".

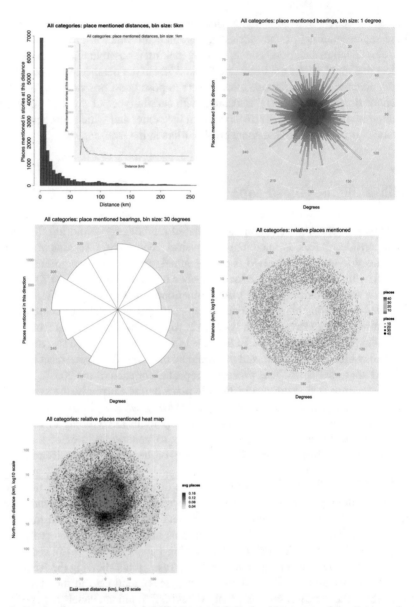

Fig. 1 Overlapping panels (**a**) and (**b**) in the *top-left* display histograms of the distances from the places told (PT) to the places mentioned (PM) in all 20,431 ETK stories using 5- and 1-km-wide bins, respectively. Panels (**c**) (*top-right*) and (**d**) (*middle-left*) display aggregate "wind rose" histograms of the directions from the PTs to the PMs with one- and 30° binning, respectively. The length of each ray corresponds to the total number of place references in that direction. Panel (**e**) (second panel in the *right column*) plots the relative distances and directions to all PMs in the corpus, plotted as points with larger point sizes representing overlapping relative positions. Panel (**f**) (the *last panel*) is a heat map plot with darker shading of points indicating multiple overlapping PM references. The *shaded regions* represent the aggregate "heat" kernels around each point; *darker shades* indicate greater concentrations of PMs at this distance and direction from the PTs

(e), the second panel in the right column of Fig. 1); and (4) directional "heat maps" showing the concentration and relative distance and direction from center of the places mentioned using a standard Gaussian kernel to represent the "reach" and "decay" of stories across geographic distances and directions (Panel (f) at the bottom of Fig. 1).[8] Due to the strong representation of PT and the rapid drop-off in PM as distance from PT increases in the latter two types of visualizations, we found it helpful to plot these graphs on logarithmic scales in which concentric rings drawn at equal intervals correspond to distances of 1, 10, and 100 km from the center. This signal-processing step allows directional patterns at close ranges to be more readily discerned while simultaneously aggregating and highlighting phenomena that occur at progressively greater distances. Our default "heat map" visualization also masks out all PT = PM references [e.g., Fig. 1 Panel (e)], which helps to prevent the relatively high incidence of reflexive place references from "washing out" other interesting patterns that the shaded regions of the heat map may expose. Finally, an interactive element to these visualizations on the website allows researchers to drill down to the underlying stories.

To derive a baseline, we calculated the PT-to-PM distances for the corpus in its entirety, using standard formulae for computing great-circle distance and bearing (forward azimuth) over the surface of the earth (Vincenty 1975). The first set of visualizations reveals that places mentioned tend to be relatively close to the storytellers' homes (<10 km) and that the directionality of the places mentioned are evenly distributed throughout the landscape (Fig. 1). In earlier work, we determined that storytellers tell stories about places that are close to where they live, and largely coterminous with the geographic extent of their social and economic activities, with perturbations in these distributions related to gender and occupation (Tangherlini 2010). In the aggregate representation of the entire corpus, the pattern of telling stories about things situated close to home is more pronounced. There are 14,254 references to a place that is physically distinct from the collection place, compared to 5,058 references to the collection place. Considering all of these, the mean distance is 31.45 km and the median is 18.65 km. Considering only references to places beyond the collection place (PT), the mean distance is 42.787 km and the median is 71.56 km. More notably, when PT's are excluded, approximately half (50.6 %) of PM's fall within ten kilometers of the place told, with a range of 416 km, which is similar to the geographic extent of Denmark. The directional histograms (Fig. 1c, d) suggest that, in the aggregate, no direction is preferred over any other direction. This point is driven home by the heat map in Fig. 1e. In other words, no matter the direction in which one looks, extending out to 150 km, there is bound to be a story—and quite possibly a fairly threatening creature. Consequently, when the stories are broken out by topic, divergences from this very even distribution suggest a consensus about the conceptual directionality and distance of that topic.

---

[8]It is an open question whether story reach across geographic distance, where story reach is the extent to which people in an area would be expected to recognize the story, is best modeled using a Gaussian kernel.

There are several phenomena worth considering in these visualizations of the entire corpus. In regards to the directional point distribution diagrams, there is a considerable gap between the center (PT) and the first significant occurrences of places mentioned, which begin to appear with considerable frequency at a threshold of 5 km and essentially disappear at approximately 250 km. The donut-like shape of the point distribution plot (Fig. 1f) suggests an intriguing general mapping of the landscape of belief. The lack of places beyond 250 km is readily understood: given both physical limitations on travel through the nineteenth century and the political and linguistic boundaries of Denmark, things that happened at such a great distance would have had little cultural relevance in the day-to-day storytelling of the rural population. Distant places within the country would be recalibrated to the local environment, a process theorized as the "localization" of legend in performance and reflected by this point plot (van Sydow 2013b; Tangherlini 2015: 11). At the same time, there would be little reason to tell stories about events occurring beyond political and linguistic boundaries, or within or beyond significant physical barriers such as the ocean where place names usually did not exist.

The gap between zero and 5 km (Fig. 1b) is at first a bit harder to understand, yet it may give us a clear indication of what was considered "local" in nineteenth century Danish tradition. The places-mentioned-by-distance frequency graph in Fig. 1 illustrates the sharp spike in place names at approximately 5 km, which is echoed in all directions in the point distribution plot. This surprising gap reflects two related phenomena: First, a storyteller's interlocutors would be expected to know local landmarks and local place names so well that they would not need to be mentioned by name. Second, a lot of local names are not easily resolved to geographic coordinates. In a story such as the following, told in Hestbæk (PT), the limitations of our current system become apparent:

> De har set, at bjærgfolkene har lagt deres sølv og guld ud at soles, der ude på Hvirvelbakke. En gammel mand, Kræn Boel, kom fra Ytrup en aften, og han så, de kom kjørende med fire heste for en kane og flere hunde kom farende rundt omkring fra Hvirvelbakke op gjennem den dal, der går op ad Ytrup (DS I 643)
>   [They had seen that the hidden folk had laid out their gold and silver to sun out there on Hvirvelbakke. An old man, Kræn Boel came from Ytrup one evening and he saw them come driving with four horses hitched to a sleigh and lots of dogs came running about from Hvirvelbakke and up through that dale that goes up towards Ytrup.]

For this story, only the place referents that can be resolved to historical gazetteers are encoded: Hestbæk (PT) and Ytrup (PM), which lies 2.5 km southwest of Hestbæk. The encoding misses the reference to Hvirvelbakke, a hill to the north east of Ytrup, and halfway to Hestbæk. Even though Hvirvelbakke appears labeled on the high board survey maps from 1882 to1883, it does not appear in the historical gazetteer. The referenced dale is even harder to resolve, given its lack of name. Consequently, for certain phenomena, the calculations and visualizations present a biased view, since many local place referents are missing. Future refinements that detect and resolve these fine-grained place names, and incorporate both local knowledge and non-specific place and topographical referents, should provide greater detail related to the directionality and relative distances ascribed to folk belief elements. Yet, even in the absence of this higher degree of precision, the geographic representations shown here still provide useful indications of the trends in tradition regarding the distribution of aspects of folk belief in the landscape (Tangherlini 2015: 75).

Ground truth is notoriously difficult to establish from Humanities data. In this case, there is no pre-determined "right" answer nor is there any clear method to determine whether the aggregate directionality or relative distance of a story topic is "correct" as measured against some standard. Of course, "correctness" may be the wrong evaluative criterion. These tools, and other tools from the *Folklore Macroscope*, are designed to reveal patterns in the data which can then be used to support hypotheses or drive the development of research questions. Given this more open-ended analytical regime, a reasonable objective for a spatial-semantic analysis tool like *GhostScope* is to highlight features in the data set that may merit further inquiry, such that they appear more prominent than other features which have lower (but still non-zero) potential. Therefore, despite the lack of unassailable ground truth, a suitable and methodologically productive approach is to compare the distributions of specific topics in the storyteller's perceived surroundings to the aggregate distribution of all topics in the environment. In particular, there are several topics that can function as benchmarks to indicate whether the approach described above provides plausible results, namely, topics that relate to a strong geographic delimiter, such as the ocean, or those that have been shown in other scholarship to have a strong geographic correlation. It is also possible to quantify and compare these differences numerically, such as via a spatial covariance test (customized for the radial, aggregated "radar-scope" environment considered here), but given the subjective nature of this endeavor, we find visual comparison of the graphs generated via *GhostScope* to be sufficient.

In the first case, we looked at stories related to strandings, a fairly common phenomenon along the north and west coasts of Jutland where the capricious nature of the North Atlantic coupled with difficult currents and hidden obstacles posed significant danger to transport and fishing vessels alike up through the twentieth century. Because of laws governing salvage, strandings presented an opportunity for people to acquire unexpected fortunes yet also set the stage for potential murders, hauntings, and encounters with threatening foreigners. A typical stranding story, collected in Staby (PT), reads as follows:

> Et norsk skib strandede en bælgmørk nat der ude i nærheden af Torsminde. En af mandskabet blev med en bølge slået ind i havstokken. Han greb nu med fingrene ned i sandet for at holde sig fast, men det gav sig, og han blev ved at bore fingrene ned så længe, til fingerenderne var helt blodige. Den ene bølge kom og vilde drage ham med sig efter den anden, og til sidst var han nær ved at miste besindelsen. Men så kom han i tanker om sin lommekniv, fik den op og huggede den ned i sandet. Endelig nåede han op. Så reddede han tre andre. Der var slet ingen ved stranden, og de kunde ikke finde op gjennem bjærgene. Lidt efter traf de en anden lille flok, der også var kommen i land fra skibet. Endelig traf de en forladt fiskerbåd, og så måtte de blive der i det våde töj til dag. (DS IV 1645)
>
> [A Norwegian ship stranded one pitch-black night out there near Torsminde. One of the crew was tossed up onto the beach by a wave. He grabbed down into the sand with his hands to hold on, but it gave way, and he continued boring his fingers down so far until his

fingertips were completely bloody. Then one wave after another washed over him and was
going to drag him out, and he was just about to lose consciousness. Then he remembered his
pocket knife, got it out and stabbed it down into the sand. Finally he was able to pull himself
up. Then he saved three others. There was nobody there at the beach and they couldn't find
a way up through the cliffs. A little later they found another small group that had also made
it ashore from the ship. Finally, they discovered an abandoned fishing boat and then they
had to stay there that night in their wet clothes.]

Here, the resolvable places are Torsminde (PM), Staby (PT) but not the somewhat
imprecise Norwegian origin of the boat. The beach, presumably near Torsminde, is
not easily resolved.

Two aspects related to distance and direction for stories about strandings should
be observable if this overall model is to be considered useful. First, since the greatest
danger for stranding occurred along the west and northwest coasts of Jutland, there
should be a strong bias toward these directions in the aggregate places mentioned.
Similarly, there should be a strong negative bias against the east and south since, in
Denmark, those directions were either characterized by overland routes or fairly safe
navigable waters that were not likely to produce strandings. Both of these biases are
found in the data and clearly represented in the directional histogram visualizations
(Fig. 2b, c). In addition, a heat map representation (Fig. 2e) reveals a clear visual
correlation between these stories and the coast, with the heat map taking on a shape
reminiscent of the coast of Jutland.

Second, the relative distances represented in these stories should capture several
competing phenomena. Since people tend to tell stories about places close to where
they live, there should be the standard spike in stories at relatively close distance.
But, because these coastal areas were sparsely populated, it would make sense if
the relative distance revealed a slight bias toward greater "local" distances since, for
people living in these coastal villages, the concept of "local" probably extended
to greater distances than for people living in more populated areas in eastern
Jutland. This phenomenon is also captured in the graphs, with stories appearing
with higher frequency in the range of 10–20 km as well (Fig. 2a). Finally, since
most non-coastal Danes were aware that ships run aground and since they also knew
that the places where ships stranded were along the northwest coast, when these
people told stories about these events, one would expect them to situate strandings
at distances likely to be in the 50–100 km range. These latter two phenomena
are also reflected in the data, as shown in the heat map graph (Fig. 2e) and the
directional point distribution diagram for strandings (Fig. 2d). Of particular interest
in these diagrams is the concentration of stories at 315° at a distance of 100 km,
suggesting that people who lived away from the coast and were thus less likely to
be familiar with west coast place names chose a "most likely place" for strandings
and assumed that their local audience would accept the specific place referent as
a necessary stand-in for the broader conceptual category of the dangerous west
coast.

A slightly different approach is applied to stories about plagues and other
illnesses. Earlier work suggests that disease, in particular the plague, came to
Denmark along the two most important trading routes: boats arriving from the west,

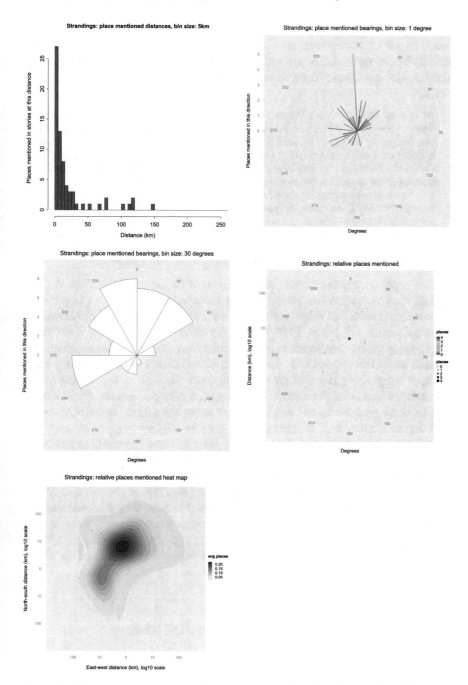

Fig. 2 The "Strandings" category. Panel (**a**) contains a histogram of the distances from PTs to PMs for the category with distances grouped into 5-km bins. Panels (**b**) and (**c**) are wind-rose histograms binned into 1- and 30° intervals. Again, the length of each "ray" corresponding to the number of place references in that direction. In panel (**d**) larger points indicate multiple overlapping relative positions and panel (**e**) is a heat map for "Strandings"

and overland travel from the south (Tangherlini 1988). This real-world phenomenon was echoed in stories and, over the years, created a broad consensus of the association of these directions with disease (Tangherlini 1988). A story told by Kristen Meldgård from Tapdrup recounts the following chaotic outbreak of the plague:

> En kone fra Nedre-Hornbæk var med hendes to sönner gået til Randers. Da de gik hjem igjen, og hun tråj ud af Hvide-mølle vold og altså lige kom på Hornbæk ejendom, da hørte hun, det gav sig til at ringe underneden dem i jorden. Så løb en af sönnerne hjem til deres hus efter en spade, og så kastede han, hvor det ringede, og nu begyndte det at kime, og så kimede det i 3 timer. Ikke svar længe efter udbrød den sorte pest. Den udbrød i Randers først, og de mente, den var ført dertil ved skibslejlighed. (DS IV 1682)
>
> [A woman from Nedre-Hornbæk had walked with her two sons to Randers. When they were on their way home again, and just as she stepped off of Hvidemølle ramparts and had just gotten back onto Hornbæk property, she heard ringing below them in the ground. Then one of the sons ran home to get a shovel, and they dug where it was ringing, and then it began to peal, and then it pealed like that for 3 hours. Then it wasn't long before the plague broke out. It broke out in Randers first, and they believe it arrived there by ship.]

The repeated epidemics that swept Denmark in the centuries from the late middle ages to the modern period provided storytellers not only an impetus for the continued telling of these stories, but also contributed to the reinforcement of these associations (Lindow 1973-74). At the same time, the development of an increasingly robust transportation infrastructure over those centuries contributed to a reduction of this effect. Consequently, plague and illness could appear anywhere in the country—indeed one of the terrifying aspects of disease was the capricious nature of its sudden appearance and rapid spread. Since disease is not as closely tied to fixed geographic features as stories about strandings, one would expect that the association with specific directions to be slightly less pronounced. The visualizations make this quite clear, with the south and west favored in these stories, but with a noticeable representation of other directions as well (Fig. 3a). The distances between the storyteller and the plague are more difficult to parse, with a surprisingly even distribution from near to far (Fig. 3b). In the example above, Tapdrup is 33 km due west of Randers and Nedre Hornbæk. Danish legend tradition therefore proposes an amusing challenge to the conceit of Bocaccio's *Decameron* (1353), where the wealthy aristocrats escape the plague by running for the hills. In Denmark, by contrast, no matter where you turn there is no escape from the plague (not even to tell stories!).

Supernatural Creatures and Where to Find Them

Supernatural creatures are among the best known features of Danish legend tradition, and each of these creatures is, in folklore scholarship, generally thought to have a particular locus of action. Furthest from settlements and arable land one finds giants (*jætter, kæmper*), although traces of their activities encroach into arable fields and leave their mark on churches. In the woods and forests, as well as in small groves

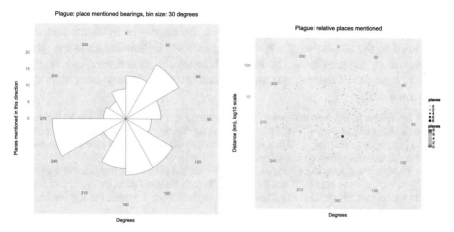

Fig. 3 "Plague"—panel (**a**) is a wind-rose histogram with 30° "slices" of the PT-to-PM bearings in all stories in the category. Panel (**b**) is a plot of the logarithmic distances and directions to all PMs from a collective, central PT in the category

of trees near streams, one finds elves (*ellefolk*) and other figures such as the Wild Huntsman (*Odins jæger*). Water spirits inhabit lakes and rivers, while merfolk live in the oceans. Basilisks and serpents generally inhabit boundary areas, such as stone walls, dikes, and embankments, while wyverns have a predilection for churches and cemeteries. The fields and, in particular the mounds that dot these fields, are home to the hidden folk or "mound dwellers" (*bjærgfolk, højfolk*). Within the confines of the farm or house, one finds the *nisse*, popularized in the *Harry Potter* series as the "house elf" and akin to the English brownie or German *kobold*. In many of these stories, the place names associated with these supernatural creatures are local in nature while other geographic referents amount to generic descriptions of landscape features, either natural or manmade. Unfortunately, *GhostScope* is unable to capture the association of these supernatural phenomena with such features.[9] Nevertheless, resolvable toponyms do help delimit areas of interaction for these creatures, as well as situate the action either close to the storyteller or at a greater distance. A story that happened "here" has a different rhetorical resonance with audiences than a story that happened "over there".[10]

Stories about giants often have an etiological component to them, relating the actions of giants either to the building of churches or strange geographic features. Unusual architectonic features of local churches, such as spires in a different style

---

[9]A planned future tool, MoundMapper, provides a rank list of geographic descriptors that co-occur with each supernatural phenomenon. These associations will be represented as a network, facilitating various graph-analytical operations. This approach to classification can be found in a broader form in Abello et al. (2012).

[10]In studies of the fairy tale, Holbek notes that physical distance may be a narrative representation of psychological distance and, in turn, the immediacy of the threat (Holbek 1987).

than the body of church, or landscape features such as regularly shaped pools, deep clefts in rocks or cliffs, and large stones in the middle of fields are frequently attributed to the actions of giants. The large stones, technically known as glacial erratics, are a common feature in the Danish landscape. In narrative, the seemingly random locations of these stones are attributed to ancient feats of strength or fights between giants:

> På Dybbøl bakke ligger en meget stor sten, tidligere har den været så stor, at man har kunnet vende med en vogn oven på den; nu er den betydelig mindre. Om denne sten går sagnet, at en kjæmpepige, som boede på Als, en dag slyngede den efter sin kjæreste i sit hårbånd; men den faldt temmelig tidlig og ramte ham ikke (DS III 95)
>
> There is a really big stone lying on Dybbøl hill. It used to be so big that one could turn a wagon around on it, but now it is much smaller. There's a legend about that stone that says that a giant girl, who lived in Als, slung it at her boyfriend one day using her hair band, but it fell short and didn't hit him.

Worth noting is that the actions of giants are usually set in the distant past. In storytelling, physical distance can act as a metaphor for temporal distance, and may help explain the general association of giants with greater distances (Holbek 1987). Giants, of course, were not considered by nineteenth century Danes to be a current threat. Giants had no invisibility powers such as those of the hidden folk, and the relatively flat and evenly settled Danish landscape made it impossible for them to hide.[11] Consequently, in Denmark, there is no compelling need to keep giants at a distance—their threat is gone. All that remains is their aftermath, which can now be read in the local landscape. In the point distribution diagram, many of the stories cluster at a distance of 10–20 km—about the distance a giant can throw, if the above story is to be trusted—while in the heat map, one discovers a weak southwest-northeast orientation [Fig. 4, Panels (a)–(c)]. This fairly weak effect may have an explanation as this axis follows the direction of the retreat of glaciers across Denmark. Consequently, it would also be along this line that one would most likely encounter glacial features such as erratics that are, in storytelling, attributed to giants.

The visualizations of the elves (or at least the elf stories) are perhaps a bit more revealing, clustering just outside the reach of the "local" and inhabiting a dangerous circle that defines the edge of the civilized [Fig. 5, Panels (a)–(d)]. As seductive nature spirits, elves attack the very foundations of human society, undermining the local economy by luring people away from their families and work, and striking at human fertility through their attempts to seduce. Rasmus Dyrholm tells one such story:

> Der var en ellekjælling her nede i Skagelund-skoven nede ved Råby kjær, hun sad og gav et barn patterne inde i skoven, og hun var hul æbag ligesom et bagtrug. Der kommer så én gangende og får dether at se. De kaldte ham den tyske Krage, for han var noget tysk i snakken, og han var fra Dalsgården. Hun smider barnet og af sted efter kam. Han stak jo af,

---

[11] This contrasts sharply with the rugged landscape of Norway that offers trolls, the Danish giants' cousins, huge expanses of unsettled territory, tall mountains, and dense forests in which to hide.

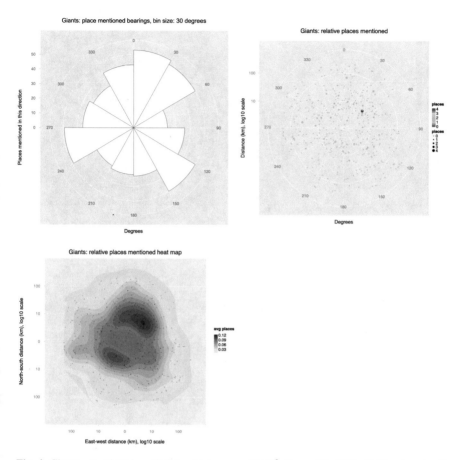

Fig. 4 Giants—panel (**a**) is a wind-rose histogram with 30° slices of the PT-to-PM bearings in all stories in the category. Panel (**b**) plots distances and directions to all places mentioned (PM) and panel (**c**) is the corresponding heat map

da hun smed knægten, og han kom endda lykkelig hjem, så hun fik ikke fat på ham. (DS II 54)

[There was an elf hag down here in Skagelund forest down near Råby marsh, she sat and nursed a child out in the forest, and her back was just as hollow as a dough trough. Then someone comes walking along and sees this. They called him the German crow, since he had a bit of a German accent, and he came from Dalsgården. She throws the child away and runs after him. He took off when she threw the little boy, and he manages to get home safely, so she didn't get hold of him.]

The point distribution diagram illustrates the locus of the elves best. Hovering on either side of the 10 km mark, yet equally dispersed in all directions, the diagram shows that elves are pervasive in the landscape, yet associated with areas just beyond the immediate. The PM histogram bears this observation out, with concentrations in the 5–15 km range. Intriguingly, mentions of elves stretch far to the horizon, with

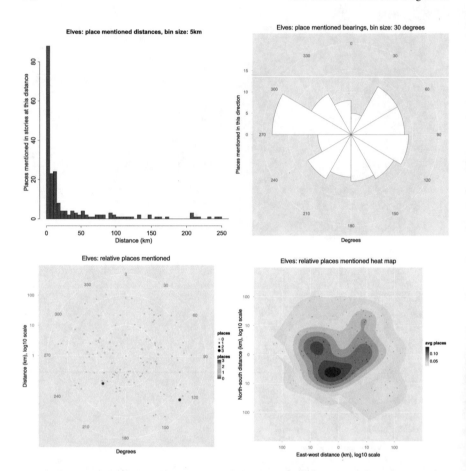

Fig. 5 Elves—panel (**a**) gives the PT-to-PM histogram, with distances in 5-km bins. Panel (**b**) is a wind rose with 30° binning. Panel (**c**) represents distances and directions to all PMs and panel (**d**) is the corresponding heat map

some stories mentioning places several hundred kilometers distant from the places where they were told.

Although less pronounced than the distance associations, there is a minor directional association for elves toward the west. Up through the nineteenth century, western Jutland was marked by large areas of non-arable land. Parts of it were covered by heath (*heden*), while large expanses in the northwest were covered by bog (*store vildmosen*). Small forests dotted the area and were, by law, off limits to most of the rural population. In short, the west was widely associated with "the wild" and the uncultivated; it was imagined as a frontier, with all of the dangers associated with such regions. The slight western bias in stories about elves bears out this notion, and doubtlessly contributed to the ongoing development of the consensus of the west as wild, untamed, and potentially dangerous.

Closer to the space of everyday life, one encounters the mound dwellers or hidden folk, the most ubiquitous of supernatural creatures in Danish legend tradition, with mentions in 2,404 stories or approximately 7 % of the target corpus. They were closely tied to cultivated land, living quite literally in it, with dwellings inside the mounds and hillocks that dot the Danish countryside. Living a parallel existence to humans, they had families and farms, partook in the normal activities of everyday life, engaged in trade, brought goods to market, and had lively social lives that included banquets, dances, and other life-cycle and seasonal festivals. Apart from being largely invisible and living underground, the main delimiter between humans and mound dwellers was that humans were Christian while supernatural beings clearly were not. At times helpful, at times threatening, stories about mound dwellers capture well the insecurities that accompanied the remarkable changes in Danish agricultural organization of the nineteenth century (Tangherlini 2013a, b).

Mound dwellers appear with almost equal frequency in all directions, yet they occupy an area slightly closer to the storytellers than the elves (Fig. 6). Interestingly, the range for the mound dwellers is far more constrained than that of elves and giants, with an area narrowly concentrated on either side of the 10 km ring, approximately between 3 and 12 km. Although the complete range stretches beyond 100 km, the majority of the stories (66 %) fall well inside the first 20 km. The shifting organization of Danish farms may help explain this phenomenon.

Prior to the land reforms that marked the late eighteenth and nineteenth centuries, fields generally encircled villages, and distances between villages could be quite large. Farmers would move from field to field with their animals and plows to till narrow strips of much larger communally farmed fields. The reapportionment of land that began in the late eighteenth century consolidated these widely distributed strips into single fields, and thus refigured large parts of rural Denmark. To take

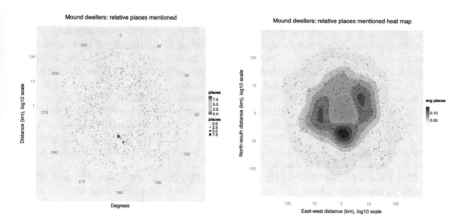

Fig. 6 Mound Dwellers—panel (**a**) gives the distances and directions to all PMs in the category while panel (**b**) gives the corresponding heat map

advantage of the efficiencies created by these newly reorganized fields, farms and houses were physically moved from centralized settlements onto their aggregated fields. This restructuring led to a marked change in population distribution and settlement patterns. While distances between farms were reduced, population densities were also reduced. Even though the land reforms gave people more control over their own land, it also forced them into a position of greater responsibility. A farm's fortunes could rise and fall quickly based on harvests, unpredictable weather, and the caprices of reapportionment that resulted in some farms receiving unusually good land and others receiving unusually bad land.

While reapportionment changed where people lived, it did nothing to where the mound folk lived. They remained where they had always been, living out in the fields that separated the newly independent farms. The liminal nature of the fields—necessary to survival, but situated between the "inside" of the actual farm, and the "outside" realm of heath, swamps, woods and forests—is captured by the liminal nature of the mound dwellers who straddle both the good and the bad, existing, to borrow Turner's formulation, "betwixt and between" (Turner 1967). Mirroring the agricultural landscape, the liminal mound dweller is situated in storytelling in a middle range of distances—close enough to be part of the everyday, yet far enough to be outside the safe confines of the farm itself.

The *nisse* is perhaps the most "inside" of the Danish supernatural creatures (Lindow 1975). Unfortunately, because these domestic creatures are so tied to the farm, *GhostScope* is unable to properly capture many of the geographic referents in these stories such as barns, stalls, haylofts and courtyards. There are, however, several exceptions to this general rule. The *nisse*, while closely associated with a specific farm, occasionally leaves it, either as punishment to the farmer for mistreatment or, as in the story below, to steal from other farms:

Der var en Gaardbo [=*nisse*] i Stempegaard i Børglum Sogn—det var da i den ene af de to Gaarde af det Navn, og han havde en rød Ko at passe, og den skaffede han Foder til. Nu var det en Bestemmelse, at han hver Aften skulde have søde Grød og Smør i Grøden, og det fik han jo i flere Aar. Men saa var det en Aften, Konen vilde have ham lidt til Bedste, og saa kom hun Smør i Bunden af Fadet og Grød ovenpaa. Naa, Gaardboen han bliver jo saa gal over, at der var ikke Smør i Grøden, og han gaar ud i Kostalden og slaar den røde Ko ihjel. Derefter gaar han ind og fryder sig over, hvad han har gjort. Men da han havde spist Grøden, finder han Smørret paa Bunden af Fadet og fortryder det nu lige godt. Nu vidste han, at der ovre i Vollerup—saadan kaldes nogle Steder, der ligger nord derfra i Em Sogn, der stod en rød Ko, akkurat Magen til den, han havde slagen ihjel. Saa tog han den døde Ko paa Nakken og rejste over en Hede, der kaldes Skrølløs Hede imellem Stenbækgaard og Vollerup, der var fuld af Sten og Skidt. Han trækkede den levende Ko ud og bar den døde ind og satte i Steden for den anden, og saa rejste han tilbage igjen til Stempegaard med Koen og satte den ind. Andendagen gik han og snakkede med sig selv: "Aaja, Skrølløs Heje wa lang, å dæj røø Ko wa tång, åå mi Nøg, hu bor a." Han gik jo og ynkede sig selv, det havde gjort ondt i hans Ryg, og saa hylrede han over sig selv (DSnr II B 212)

[There was a gaardbo (nisse) in Stempegaard in Børglum parish—it was in one of the two farms that had that name, and he had a red cow he was supposed to take care of, and he procured food for it. Now there was a condition that he was to have sweet porridge with butter in it every evening and he got this for many years. But then one evening the farm wife wanted to play a little trick on him and so she put the butter on the bottom of the bowl and the porridge on top of it. Well, the gaardbo gets so angry that there wasn't butter in the porridge and he goes out into the cowshed and kills the red cow. Then he goes back inside and is delighted with what he's done. But when he'd eaten the porridge, he finds the butter at the bottom of the bowl and he now regrets what he's done. Now he knew that over in Vollerup—that's what those places north in Em parish are called—there was a red cow there exactly like the one that he'd killed. Then he puts the dead cow up on his shoulders and went over the heath that's called Skrølløs heath between Stenbæk farm and Vollerup, it was filled with stones and other junk. He pulls the live cow out and carries the dead cow in and puts it in place of the other one, and then he went back again to Stempegaard with the live cow and put it in the barn. The next day he went and talked with himself, "Oh, yeah, Skrølløs heath was long, and the red cow heavy, oh my back, I carried it." He went and complained, his back really hurt, and he went about whining.]

The theft from other farms reveals a central characteristic of the *nisse*, namely his association with personal success and wealth (Lindow 1975). In contrast to the earlier organization of farms as part of larger manor farms, in which individual farmer's decisions were subordinate to those of the *fælleskab*, or communal decision making body, the newly emerging economic order was more in line with free market thinking, where even the smallest smallholder had to behave like a market actor. As the system evolved, Danish farmers became more attuned to the vicissitudes of the market, and the uneven distribution of wealth and success. A particularly successful farmer was often thought to have a helpful *nisse*, or perhaps a larcenous *nisse* not averse to a forceful redistribution of local wealth as above.

Even acknowledging that most *nisse* activity is not adequately represented by *GhostScope*, the stories that the system captures reveal several interesting aspects of *nisse* behavior. First, in comparison to the distance distributions for elves and hidden folk, the *nisser* are a much "closer in" phenomenon, best illustrated in the point distribution diagram (Fig. 7b). Although the range extends well beyond 100 km, recognizing that *nisser* are found throughout Denmark, the mentions that fall well within the 10 km circle are particularly noteworthy. This concentration most likely captures two simultaneous phenomena linked to the strong economic nature of the *nisse*: the suspicion that a neighbor's success is due to the intervention of a *nisse* as opposed to any particular farming prowess, and the movement of *nisse* through the fields to other farms for their larcenous exploits. There is also a very strong directional association in these stories with the south, best seen in the binned directional histograms and the heat maps. A possible explanation for this bias may be the link between the shifting focus of most farmers' and smallholders' economic fortunes on local trade toward more regional and even international trade. In Denmark, the main trade routes run toward the south, with roadways and train lines bringing goods to harbors and market towns, and further afield to the international markets of continental Europe and England. In that interpretation, the *nisse*'s connection to economic success echoes the actual economic success of farmers able to move their goods south and on to more lucrative markets.

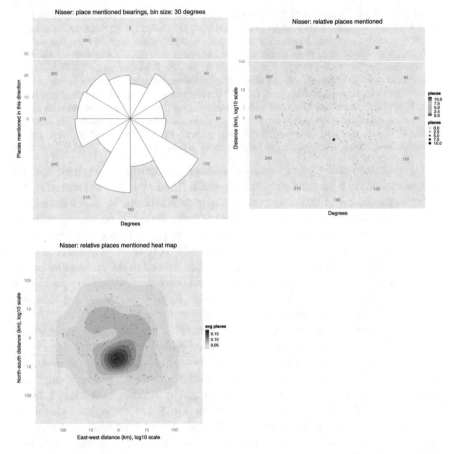

Fig. 7 Nisser (House Elves)—panel (**a**) represents a wind-rose histogram with 30° slices of PT-to-PM bearings in the category. Panel (**b**) gives distances and directions to all PMs and panel (**c**) is the corresponding heat map

Riches in the Landscape: Treasure, Robbers, and the Community

The association between fortune and the land is not limited to stories about supernatural creatures. Perhaps the clearest associations between land and wealth involve stories about buried treasure.[12] Buried treasure distributions capture the competing

---

[12]The initial goals of this project were, in fact, to find buried hordes of treasure. Fed up by low academic pay, we speculated that these legends might actually hold actionable information about where treasure could be found, as opposed to simple metaphoric representations of the connection between wealth and land. When we discovered that we would only be able to claim a small percentage of any discovered treasure given restrictions placed on us by the university, the Danish

ideas communicated by these stories: On the one hand, finding a treasure buried in the ground would be an incredible stroke of fortune, and represent a considerable windfall for an individual, much like winning the lottery in contemporary society. On the other hand, the prescriptions that treasure hunters must follow in Danish tradition, such as not talking, challenge the very nature of community itself. In a story of an unsuccessful hunt, the potential that acquiring the treasure has for destroying the community is vividly depicted:

> Stjærnehöj ligger sydvestligst i Ulfborg plantage. Det er en meget anselig opkastet höj, og man kan se, at der har været kastet meget i den, men udbyttet har kun været en enkelt stenøkse. Der bliver ellers fortalt, at der gjemmes en skat i den. En gang vilde mændene i Lystbæk-gårdene prøve på at finde den og var godt i med arbejdet. De var komne så vidt, at de så randen af en stor kjedel; men da blev en af dem vaer, at deres gårde stod i lys lue. De styrtede da i fuldt løb ned af höjen, men opdagede snart, at det kun var øjenforblændelse. Skatten var dog med det samme forsvunden. (DS III 2296)
>
> [Stjærnehøj (Star mound) is in the southwestern end of Ulfborg orchard. It's a very visibly raised mound, and one can see that a lot of people have dug in it, but the yield has only been a single stone axe. It was said, otherwise, that there is a treasure hidden in it. One time the men from the Lysbæk farms wanted to try and find it and were well underway with the work. They had gotten so far that they saw the rim of a giant kettle; but then one of them noticed that their farms were ablaze. They stumbled from the mound running as fast as they could, but soon they discovered that it was simply an illusion. The treasure had immediately disappeared however.]

Here, the choice is clear: either acquire the treasure or save the community. While Denmark's emerging economic organization may have encouraged individual initiative in contrast to community solidarity, tradition values community more than anything else. Under the earlier manorial system, the prevalent economic ethos was one akin to George Foster's concept of "limited good" (Foster 1965; Lindow 1982). In contrast, the economic realities of the late nineteenth century required people to operate with an optimistic "unlimited good" model of economic behavior (Tangherlini 1988). Yet, for most people, the potentially unlimited good of this new open market must have appeared tantalizingly just out of reach. The visualization of distance related to these stories bears this out, with buried treasures inhabiting a space remarkably similar to that of the mound dwellers, appearing in all directions and clustered very close to the 10 km radius (Fig. 8). The implication is clear—buried treasure is all around, hidden out in the fields, but it is, for all intents and purposes, impossible to acquire. If one does acquire it, it may come at great cost to the community. Indeed, the few times treasure hunters succeed, they invariably use the windfall to build a church, the clearest icon of community in rural nineteenth century Denmark, sidestepping the socially destructive nature of personal wealth in a frail economy.

While treasure hunters inevitably return to their communities without any financial gains, robbers deliberately cut themselves off from the community, hiding in forests, woods, and mounds, similar to where elves and mound dwellers live. The threat robbers pose to the community is also economic in nature, with a clear

---

government, and our funders, we reluctantly abandoned this far more practical application of our technology.

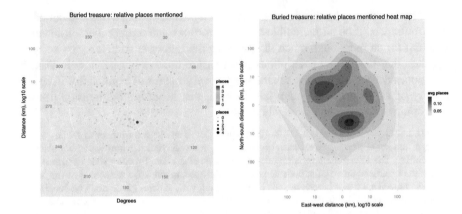

Fig. 8 Buried treasure—panel (**a**) gives relative distances and directions to all PMs in the category while panel (**b**) gives the corresponding heat map

focus on a forceful redistribution of wealth. This redistribution is not for any noble ends, but rather purely for personal enrichment. Robbers are most easily defeated by community action:

> Der er en höj lidt vesten for Kratgården i Ugilt, der kaldes Daniels höj. Her i den skal have boet en röver, som hed Daniel. Linderumgårds karle gik op og havde lidt selskab med ham, og så fandt de på en aften for løjer at bagbinde hverandre, og hvem der selv kunde rejse dem, når de var bundne, de var de villeste og de kraftigste. Karlene lod dem først binde, og de prøvede at rejse dem, og sådan blev de bundne og løste mand for mand. Siden blev han [Daniel] bunden, men så vilde de ikke løse ham igjen. De kaldte forstærkning til fra gården, og så blev han taget til fange. Han bad dem, om de ikke vilde løse ham, indtil han fik slugt det, han havde i lommen. Da var det et barnehjærte. Han havde allerede slugt de to, og så havde han den tro, at når han kunde få 3 ufødte drengebarns hjærter at spise, så skulde ingen kunne overvinde ham. Men de vilde ikke løse ham. Han blev henrettet og efter sigende begravet på kirkegården, hvor hans gravsted er på sydöst hjörne. Der har været en lille rund forhöjning, som de kaldte Daniels grav. (DS IV 1591)
>
> [There was a mound a little bit to the west of Kratgården in Ugilt, called Daniel's mound. A robber who was called Daniel was supposed to have lived in it. The farmhands from Linderumgård went up and kept company with him a bit, and then they decided one night for fun to tie each others' legs up and whoever could get up while they were tied up was the strongest and most ferocious. The farmhands let themselves be bound first and they tried to get up, and they were tied up and let go like that, one after the other. Finally, Daniel was tied up, but then they wouldn't let him go again. He asked them to untie him until he finished swallowing what he had in his pocket. It was a child's heart. He'd already swallowed two and he believed that if he could eat three unborn boy baby's hearts, then no one would be able to defeat him. But they wouldn't untie him. He was executed and, according to what they say, buried in the cemetery where his grave is in the southeast corner. There was a little round mound there they called Daniel's grave.]

Just like elves, robbers threaten all aspects of the community, their menace perhaps best encapsulated by their propensity for capturing babies and eating their hearts. In the directional and distributional visualizations, storytellers place them at a distance similar to that of the elves, in a broad band between 10 and 100 km

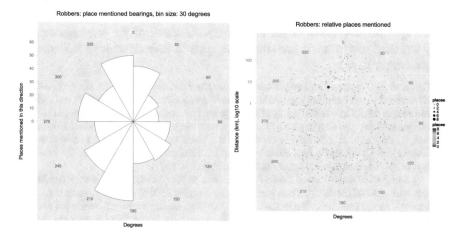

Fig. 9 Robbers—panel (**a**) is the wind-rose histogram with 30° slices and panel (**b**) gives the corresponding distances and directions

(Fig. 9). Occasional forays bring them closer to the center where they threaten the communities they have eschewed. The directional distribution reveals an intriguing pattern of a strong north–south axis, which follows the main trade axis in Denmark. This north–south orientation aligns well with the economic nature of this threat, and the emerging association of economic activity with the increasingly vital trade networks connecting Denmark with large international markets in Europe. There is also a clear positive directional bias that situates robbers more toward the west, a location that further aligns them with elves, and contributes to the conceptualization of the west as "wild".

Witches, Cunning Folk, and the Nineteenth-Century Danish "Threat Matrix"

Two of the largest categories of stories in the target corpus are those about witches and cunning folk, comprising 2558 (7.2%) and 1783 (5%) of the total corpus respectively.[13] If legend tradition is to be believed, witches were the scourge of rural life. They undermined economic activity every chance they got, stealing milk, destroying farm equipment, and killing animals, while they spread sickness and

---

[13]These counts are most likely low given ETK's somewhat problematic topic categorizations—numerous stories of witches are included in the category "About the Devil and being in league with him," while many stories of cunning folk are placed in the category "Ministers". For alternative methods for classification, see Abello et al. (2012), Broadwell and Tangherlini (2015), and Broadwell et al. (2015).

death. Drawing their power from Satan himself, they represented a powerful threat to the spiritual well-being of the community. Residents of the local community, they recruited new witches from the young girls in their families or their own or neighboring villages:

En mand gik og harvede på marken, og en tøs, han havde, gik ved siden af og vogtede hövderne. Så kom hun til at sige, dede hun kunde nok malke af en harvetænd. Ja, da vilde han nok se det, så kunde hun malke af den røde ko. Hun malkede også, men så sagde hun, hun var ræd for, den skulde gå til. "Ja, det kan være det samme", siger manden "Koen er min, og døer den, skal du ikke bryde dig om det." Hun malkede nu, så det pråsede. "Nu kommer der bløw", sagde hun. Siden vilde han have besked af tøsen om, hvem der havde lært hende det. Ja, det havde hendes bedstemoder. (DS VII 577)

[A man went and harrowed out on the field and a young girl who he had went alongside and took care of the cattle. Then she happened to mention that she could milk from a harrow's tooth. Well, he'd like to see that, she could go ahead and milk from the red cow. She milked too, but then she said that she was afraid that it would die. "That doesn't matter," says the man, "The cow is mine and if it dies, don't worry about it." She milked now so that it foamed. "Now blood is coming," she said. After that he wanted the girl to tell him who had taught her how to do that. Well, her grandmother had.]

Much like the *nisse*, witches were an "inside" threat, a fact that is captured in the distance distribution visualizations, with many points appearing well within the 10 km radius in the point distribution plot (Fig. 10). The range of these stories is very large, attributable to the many stories that situate the "Witches' Sabbath" in remote locations, often in the far north (the cathedral at Tromsø, Norway, though not mapped in the data set considered here, is one of the most popular such places in Danish legends). Unfortunately, the coarse grained nature of the classification, coupled to the localized nature of the vocabulary associated with witches, make it difficult to provide a clearer analysis of the geographic locus of the witches and the different types of activities in which they engaged. Despite that, in the aggregate, there is a strong eastern bias in these stories, precisely the opposite of the bias

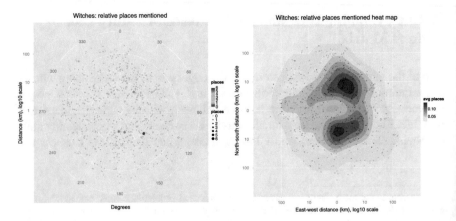

Fig. 10 Witches—panel (**a**) gives distances and directions to PMs in the "witch" category while panel (**b**) gives the associated heat map

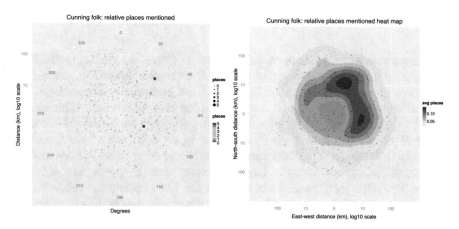

Fig. 11 Cunning folk—panel (**a**) records distances and directions to PM and panel (**b**) give the heat map for the category

for elves, robbers, thieves and murderers. While those latter threats appear to be associated with the "wild", the threat of the witch appears more closely aligned with those of the "civilized" and the distinct threat that witches pose toward civil society, attacking it from within.

The geographic distribution of cunning folk is remarkably similar to that of witches (Fig. 11). In rural Denmark, the borderline between being accused of witchcraft and being considered a cunning person was vanishingly small (Tangherlini 2000). During the time of the witchcraft trials in Denmark (1540–1693), legal provisions eventually provided a modicum of protection from persecution for folk healers and cunning folk, reserving the title of witch for those who engaged in "maleficium", similar to the malicious curses that Thor's antagonists in Hárbardsljóð had mastered. After 1686, sanctions for witchcraft were greatly lessened, and accusing someone of witchcraft was no longer likely to result in execution. Rather, these accusations became part of the behavior of cunning folk eager to capture a greater market share, with cunning folk and their customers regularly deploying stories to cast competitors as witches while playing up their own remarkable abilities (Tangherlini 2000). Often what separated a witch from a cunning person was not so much practice as community consensus. Not surprisingly, then, the directional and distance distributions for stories about cunning folk are nearly identical to those for witches. At least in the conceptual map, witches and cunning folk are remarkably similar, only distinguishable by their desire to do bad or good.

Conclusion: Crowd-Sourced Folk Geographies

The aggregate representations of place names in *GhostScope* offers a first-level approximation of the consensus of thousands of Danish storytellers concerning the locus of various aspects of folk belief in the landscape—an example of the role of folklore as the original form of cultural crowd-sourcing. Although not tuned to generic descriptions of natural and man-made environmental features, the tool does capture well the mapping of topics and places seen from the perspective of storytellers. The concept of individual perspective is often lacking in geographic representations of large amounts of data. Here, we recognize that storytellers often see themselves and their immediate surroundings—their house, their farm, their village—as standing at the conceptual center of their experiences. By placing all of the storytellers at the center, essentially having them stand on one another's shoulders, *GhostScope* provides a useful representation of relative distances and directions. Consequently, we are able to capture how the "west" becomes aligned with particular types of threat, imagined as a "wild" frontier. The east, conceptualized as civilized, is threatened by witches who attack from within, while it is simultaneously protected by cunning folk, whose locus of activity is nearly identical. The various supernatural creatures that populate Danish folk belief are shown to have specific geographic associations, revealing the conceptual mapping of the landscape into outside, liminal, and inside areas, each threatened in different ways, and each tied to the shifting economic terrain of late nineteenth century Denmark. There are various modifications and enhancements that can be made to *GhostScope*, yet this preliminary study has shown the usefulness of this type of geographic modeling. While it may not be accurate enough to help us find buried treasure just yet, it may help us steer clear of elves, witches, robbers and thieves.

References

Abello, J., Broadwell, P., & Tangherlini, T. R. (2012). The trouble with house elves: Computational folkloristics and classification. *Communications of the ACM, 55*(7), 60–70.

Broadwell, P., Mimno, D., & Tangherlini, T. R. (2015). *The tell-tale hat: Reverse engineering a folklore expert*. DH 2015 Conference. Lausanne, Switzerland.

Broadwell, P., & Tangherlini, T. R. (2012). TrollFinder: Geo-semantic exploration of a very large corpus of Danish folklore. *Proceedings of the Workshop on Computational Models of Narrative*. Istanbul, Turkey.

Broadwell, P., & Tangherlini, T. R. (2015). ElfYelp: Geolocated topic models for pattern discovery in a large folklore corpus. *Proceedings of Digital Humanities 2014*. Lausanne Switzerland.

Chesnutt, M. (1999). The many abodes of Olrik's epic laws. *Copenhagen Folklore Notes, 7*, 7–11.

Eskeröd, A. (1947). *Årets äring: Etnologiska studier i skördens och julens tro och sed. Nordiska museets handlingar* (Vol. 26). Lund: Håkan Ohlsson.

Foster, G. (1965). Peasant society and the image of limited good. *American Anthropologist, 67*, 293–315.

Grimm, J., Grimm, W., & Grimm, H. F. (1816–1818). *Deutsche sagen*. Berlin: In der Nicolaischen Buchhandlung.

Gunnell, T. (Ed.). (2008). *Legends and landscape*. Reykjavik: University of Iceland Press.

Hárbarðsljóð. In Neckel, G. (Ed.) (1983). *Edda: Die Lieder des Codex Regius. Revised Hans Kuhn* (5th ed.). Heidelberg: Winter.

Holbek, B. (1987). *Interpretation of fairy tales*. Helsinki: Suomalainen Tiedeakatemia. FF Communications 239.

Kristensen, E. T. (1892–1908). *Danske sagn som de har lydt i folkemunde*. Copenhagen: Nyt Nordisk Forlag. Reprinted 1980.

Krohn, K. (1926). *Die folkloristische Arbeitsmethode begründet von Julius Krohn und weitergeführt von Nordischen Forschern*. Instituttet for sammenlignende kulturforskning. Ser. B 5. Oslo: H. Aschehoug.

Lindow, J. (1973–1974). Personification and narrative structure in Scandinavian plague legends. *Arv* 29–30: 83–92.

Lindow, J. (1975). The male focus of Scandinavian household spirits. In R. Kvideland & T. Selberg (Eds.), *Papers IV: The 8th Congress for the International Society for Folk Narrative Research* (pp. 35–46). Bergen: Folkekultur.

Lindow, J. (1982). Swedish legends of buried treasure. *Journal of American Folklore, 95*, 257–279.

Lindow, J., & Tangherlini, T. R. (1995). Nordic legends and the question of identity: Introduction. *Scandinavian Studies, 67*, 1–7.

Moretti, F. (1998). *The Atlas of the European novel: 1800–1900*. London: Verso Books.

Piatti, B., Bär, H. R., Reuschel, A.-K., Hurni, L., & Cartwright, W. (2009). Mapping literature: Towards a geography of fiction. In W. Cartwright, G. Gartner, & A. Lehn (Eds.), *Cartography and art* (pp. 177–192). Berlin: Springer.

Tangherlini, T. R. (1988). Ships, fogs and traveling pairs: Plague legend-migration in Scandinavia. *Journal of American Folklore, 101*, 176–206.

Tangherlini, T. R. (2000). 'How do you know she's a witch?': Witches, cunning folk and competition in Denmark. *Western Folklore, 59*, 279–303.

Tangherlini, T. R. (2010). Legendary performances: Folklore, repertoire and mapping. *Ethnographia Europaea, 40*, 103–115.

Tangherlini, T. R. (2013a). The folklore macroscope: Challenges for a computational folkloristics: The 34th Archer Taylor memorial lecture. *Western Folklore, 72*(1), 7–27.

Tangherlini, T. R. (2013b). *Danish folktales, legends, and other stories*. Seattle: University of Washington Press.

Tangherlini, T. R. (2015). *Interpreting legend: Danish storytellers and their repertoires*. New York: Routledge. Orig. published 1994.

Tangherlini, T. R., & Broadwell, P. (2014). Sites of (re)Collection: Creating the Danish folklore nexus. *Journal of Folklore Research, 51*(2), 223–247.

Tangherlini, T. R., & Broadwell, P. (2016). WitchHunter: Tools for the geo-semantic exploration of a Danish Folklore Corpus. *Journal of American Folklore, 129*(511), 14–42.

Turner, V. (1967). *The forest of symbols: Aspects of Ndembu ritual*. Ithaca: Cornell University Press.

Turner, V. (1983). Liminal to liminoid in play, flow, and ritual: An essay in comparative symbology. In J. C. Harris & R. Park (Eds.), *Play, games and sports in cultural contexts* (pp. 123–164). Champaign: Human Kinetics Publishers.

Vincenty, T. (1975). Direct and inverse solutions of geodesics on the ellipsoid with application of nested equations. *Survey Review, 23*(176), 88–93.

van Sydow, C. W. (1948). *On the spread of tradition: In selected papers on folklore* (pp. 11–43). Copenhagen: Rosenkilde og Bagger.

Complex Networks of Words in Fables

Yurij Holovatch and Vasyl Palchykov

Abstract In this chapter we give an overview of the application of complex network theory to quantify some properties of language. Our study is based on two fables in Ukrainian, *Mykyta the Fox* and *Abu-Kasym's slippers*. It consists of two parts: the analysis of frequency-rank distributions of words and the application of complex network theory. The first part shows that the text sizes are sufficiently large to observe statistical properties. This supports their selection for the analysis of typical properties of the language networks in the second part of the chapter. In describing language as a complex network, while words are usually associated with nodes, there is more variability in the choice of links and different representations result in different networks. Here, we examine a number of such representations of the language network and perform a comparative analysis of their characteristics. Our results suggest that, irrespective of link representation, the Ukrainian language network used in the selected fables is a strongly correlated, scale-free, small world. We discuss how such empirical approaches may help to form a useful basis for a theoretical description of language evolution and how they may be used in analyses of other textual narratives.

Introduction

Applications of methods of quantitative analysis that are widely used in natural sciences gave rise to the discovery of one of the best known empirical relationships of quantitative linguistics. In its simplest form, this states that the probability to

Y. Holovatch
Institute for Condensed Matter Physics, National Academy of Sciences of Ukraine, 79011 Lviv, Ukraine
e-mail: hol@icmp.lviv.ua

V. Palchykov (✉)
Institute for Condensed Matter Physics, National Academy of Sciences of Ukraine, 79011 Lviv, Ukraine

Lorentz Institute for Theoretical Physics, Leiden University, 2300 RA Leiden, The Netherlands
e-mail: palchykov@icmp.lviv.ua

© Springer International Publishing Switzerland 2017
R. Kenna et al. (eds.), *Maths Meets Myths: Quantitative Approaches to Ancient Narratives*, Understanding Complex Systems,
DOI 10.1007/978-3-319-39445-9_9

randomly select the rth most frequent word in a text is r times smaller than the probability to randomly select the most frequent one (Zipf 1935, 1949). Representing the probability to randomly select the rth most frequent word as $f(r)$, this relation may be expressed by the equation

$$f(r) = A/r^{\alpha},\qquad(1)$$

where $\alpha = 1$. The empirical observation that α has the same value (namely 1) for so many natural language utterances is what is remarkable about this equation. (The quantity A is less remarkable, being just a normalization coefficient which ensures the total probability properly sums to one.) It means the distributions are "fat-tailed"—characterised by rare events happening more frequently than for normal distributions. This discovery is often attributed to its populariser, Harvard linguistics professor Zipf (1935), however, similar observations have been reported previously by Estoup (1916) and Condon (1928). Some deviations from the original form of Zipf's law (1) were later observed (Kanter and Kessler 1995; Montemuro 2001), but the fat tail of the distribution, which is a typical signature of the long-range correlations between words within corpora, remained unchanged. Moreover, analysis of vast corpora lead to a conclusion about two scaling regimes characterizing the word frequency distributions, with only the more common words (the so called kernel or core lexicon) obeying the classic Zipf law (Ferrer i Cancho and Solé 2001a; Petersen et al. 2012). While *local* organization of words within single sentences appears to be quite natural, due to the rules of grammar, long-range correlations between words are far from trivial (Kanter and Kessler 1995). Nonetheless, why such local sequences become organized *globally* may be explained by various mechanisms (Simon 1955; Li 1992).

Zipf's law and the reasons behind it provide only superficial understanding of the organization of language since the connections between the words are neglected. These connections reflect the organization of words into sentences and play key roles for transmitting information (Ferrer i Cancho and Solé 2001b). Thus, to investigate deeper structural characteristics of a language the relationships between the language units (such as words) should be taken into account. A set of those relationships combined with the corresponding language units may be naturally represented as a network or a graph. Such representations allow one to apply a number of tools to investigate the properties of the underlying system on various scales (Newman 2010).

In seeking to represent a language by a network, one may decide to relate words syntactically (Ferrer i Cancho et al. 2004, 2005; Solé 2005; Masucci and Rodgers 2006; Corominas Murtra et al. 2007; Solé et al. 2010; Barceló-Coblijn et al. 2012) or semantically (Motter et al. 2002; Sigman and Cecchi 2002; de Jesus Holanda et al. 2004; Borge-Holthoefer and Arenas 2010; Solé and Seoane 2014), for example. Alternatively one may choose to link words together based on their co-occurrence (i.e., if they appear adjacent to each other) or on their having appeared in the same sentence (Ferrer i Cancho and Solé 2001b; Caldeira et al. 2006; Holovatch and Palchykov 2007; Zhou et al. 2008; Solé and Seoane 2014). Therefore, there is no unique network representation of a language. Nonetheless many features of language networks are shared, not only among various representations, but for diverse languages as well (Ferrer i Cancho et al. 2004). The key common features of these networks include the *small world* structure (Watts 1999) and the *scale-free* topology (Albert et al. 1999), both characterizing the global picture of linking architecture. The former (the small world effect) demonstrates that the links connect the words in a specific way that makes the corresponding structures extremely compact. The latter (the scale-free topology) demonstrates that the number of links that are connected to an arbitrarily selected node in a network have extremely high levels of fluctuations. The number of links of a given node is called its *degree* and is represented by the variable k. The scale-free property is then represented mathematically by a degree distribution function $P(k)$ which has a power-law decay as follows:

$$P(k) \sim k^{-\gamma}, \tag{2}$$

in which the exponent $\gamma > 1$. (Here the symbol $\sim$ means "behaves asymptotically like", having suppressed a normalisation term.) These structural peculiarities reveal high level of system heterogeneity and have to be properly considered in investigations of the mesoscopic structure of human language (Newman 2012).

In this article we report on an analysis of the Ukrainian language through two representative texts, namely *Mykyta the Fox* and *Abu-Kasym's slippers* (Holovatch and Palchykov 2007). The former was written by the prominent Ukrainian writer Ivan Franko and the latter adapted by him into Ukrainian. For each fable, and for a combination of the two, we construct several network representations, where the links between nodes (words) are introduced in various ways based on the interaction window. The properties of the different network representations are compared to each other and features of the Ukrainian language are compared to ones of other well studied languages.

The fable *Mykyta the Fox* is based on a story about a clever Fox and his adventures in the kingdom of a Lion. *Abu-Kasym's slippers* is a story about a miserly merchant in Baghdad. One may find variants of this story in many cultures. Therefore, similar to other chapters of this book, we use written narratives for our analysis. However, here we are interested in universal properties of language rather than the structure or contents of the stories themselves. In this sense, social networks of characters and other particular features of the narrative are not relevant for this study. Instead, we are interested in the global organization of words in the texts and in their networks. The structure of the remainder of this chapter is as follows. In section "Word Appearance Statistics" we start investigating the above-mentioned examples of the Ukrainian language by analysing their frequency-rank dependencies. Verifying that Zipf's law (1) holds for the selected texts, convinces us that the samples are large enough to deliver meaningful statistical peculiarities. In section "Language Networks" the details of how to represent language by networks are described and the properties of the corresponding language networks are investigated, focusing on the small world and scale-free topologies. Conclusions are outlined in section "Conclusions".

Word Appearance Statistics

As mentioned above, the aim of our investigation is to perform a quantitative analysis of the Ukrainian language, considering two fables as its representatives. It is natural to start this analysis by verifying the validity of Zipf's law (1) for the texts that represent each of the fables separately. In particular, this will show whether the sizes of the selected texts are large enough to demonstrate the expected statistical regularities.

The original part of our investigation uses electronic versions of the two fables *Mykyta the Fox* and *Abu-Kasym's slippers*,[1] and is based on our initial analysis of these texts (Holovatch and Palchykov 2007). All the words were set to their canonical forms. The lengths (total number of words) in these texts are $N = 15{,}426$ and $N = 8002$ for *Mykyta the Fox* and *Abu-Kasym's slippers*, respectively. The corresponding vocabularies (the number of unique words) are $V = 3563$ and $V = 2392$, respectively. The text comprising a combination of the two fables has $N = 23{,}428$ separate, and $V = 4823$ unique, words. Now, let us count the number of appearance of each unique word in a text. We associate these numbers with the frequencies of appearance f, even though they differ by the normalization coefficient N. Ordering all the unique words by decreasing frequencies of their appearance, one may assign the rank variable $r = 1, 2, \ldots, V$ with each of them, such that the most frequent word has a rank $r = 1$, the rank of the second most frequent word is $r = 2$, etc. In cases where several unique words have exactly the same frequencies of appearances, they are randomly assigned sequential rank values without any preferences. The top ranked words are the ones that appear the most frequently within the texts. For the Ukrainian language these are function words that have little semantic content of their own and chiefly indicate a grammatical relationship. Words that are rather related to content of a particular text tend to have lower frequencies. Two samples of such ranking ordering are shown in Table 1.

The frequency-rank dependence for the words of *Mykyta the Fox* is shown in Fig. 1a. This dependency may be accurately described by the power-law function (1) for roughly $r > 20$. The exponent of the distribution has been estimated by using least-squares fitting with a double logarithmic scale. (In cases where ranks have the same frequencies f, a single r-value, namely its average on the logarithmic scale, has been used for fitting purposes.) The resulting exponents are close to the originally observed value; $\alpha = 1.01 \pm 0.01$ for *Mykyta the Fox*, $\alpha = 0.99 \pm 0.02$ for *Abu-Kasym's slippers* and $\alpha = 1.00 \pm 0.01$ for the combined text of the two fables. Here

---

[1]The access to the electronic versions of these texts was through the most complete internet library of Ukrainian poetry, http://poetyka.uazone.net/.

Table 1 Rank classification of words from Ivan Franko's *Mykyta the Fox* (left part of the table) and *Abu-Kasym's slippers* (right)

r	f	word (in Ukrainian)	English translation	r	f	word (in Ukrainian)	English translation
1	439	я	I	1	165	він	he
2	323	не	not	2	163	в	in
3	312	в	in	3	143	не	not
4	272	і	and	4	140	і	and
5	233	ти	you	5	128	той	those
6	222	що	that	6	125	що	that
7	214	на	on	7	125	на	on
⋮	⋮	⋮	⋮	⋮	⋮	⋮	⋮
16	140	лис	fox	12	87	капець	slipper
21	109	Микита	Mykyta	18	69	Абу-Касим	Abu-Kasym
23	98	вовк	wolf	40	28	пан	lord
25	88	цар	tsar	41	27	суддя	judge

The table shows some of the most frequently used words (prepositions, pronouns, etc., as well as nouns) in Ukrainian for each of the two fables and their English translations. Here r is the rank of the word and f is the number of times it has appeared in the text

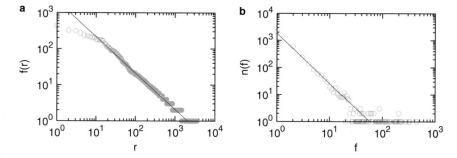

Fig. 1 Zipf laws for *Mykyta the Fox*. Panel (**a**): empirically observed frequency-rank dependency approximated by the power-law function (1) with exponent $\alpha = 1.01$ for $r > 20$. Panel (**b**): dependency of the number $n(f)$ of unique words that appeared precisely f times in the text as a function of f, with the corresponding power-law approximation from (3) with $\beta = 1.85$

the accuracies of identification of exponent α are expressed by asymptotic standard errors.

Let us stress one more specific feature of the dependency $f(r)$: the number of unique words $n(f)$ that have the same frequency f of appearance increases with r. The corresponding dependence for *Mykyta the Fox* is shown in Fig. 1b. The results of quantitative investigations (Zipf 1935, 1949) show that this dependence also follows a power law decay:

$$n(f) = B/f^{\beta}, \tag{3}$$

where B is the proportionality coefficient. The reported values of the exponent β are close to $\beta = 2$ (Zipf 1935), however, some deviations from this value, caused by specific lexicon features, have been observed [see Ferrer i Cancho (2005) and references therein]. The values of the exponent β for the texts under our investigation were estimated to be $\beta = 1.85 \pm 0.04$ for *Mykyta the Fox*, $\beta = 2.02 \pm 0.08$ for *Abu-Kasym's slippers* and $\beta = 1.80 \pm 0.04$ for the combination of the two texts. The fitting of $n(f)$ has been performed in a similar manner to that of the $f(r)$ dependence: the most general words with $r \leq 20$ have been excluded and then for each value of $n(f)$ a single value of f, its average in logarithmic scale, has been assigned.

Equation (3) is sometimes referred to as the second Zipf law [in which case Eq. (1) is called the first]. However, the two laws are not independent: Eq. (3) directly leads to (1) asymptotically. To show this, let us have a look at the rank of a word from a different point of view. The rank r of a word that has appeared exactly f times may be estimated by the number of unique words that have appeared f times or less:

$$r(f) = \sum_{f'=f}^{f_{max}} n(f'), \tag{4}$$

where f_{max} is the frequency of the most frequent word in the text. Substituting (3) into (4) and assuming that $\beta > 1$ one arrives at the inverse rank-frequency dependency for infinitely large f_{max} as

$$r(f) = \frac{B}{\beta - 1} f^{1-\beta}. \tag{5}$$

Equation (5) matches Eq. (1) provided that

$$\beta = 1 + 1/\alpha. \tag{6}$$

Thus, $\alpha = 1$ leads directly to $\beta = 2$, as observed empirically.

The power-law character of the frequency-rank distribution (1) may be explained by a number of theoretical models.[2] One of the best known is a generative Simon model (Simon 1955). This model belongs to a class of models that are based on so-called null hypotheses (Ferrer i Cancho 2005). The null hypotheses ignore some fundamental aspects of why and how the system units are used, but they often lead to qualitatively correct descriptions of system behaviour. The Simon model considers the process of text writing and uses two basic mechanisms to predict the $(n + 1)$th

---

[2]The formation of a sentence may be considered (Thurner et al. 2015) as an example of a history-dependent process that becomes more constrained as it unfolds (Corominas-Murtra et al. 2015). Recently it has been demonstrated that stochastic processes of this kind necessarily lead to Zipf's law too (Thurner et al. 2015; Corominas-Murtra et al. 2015).

word provided n words have been already written and are known: (i) the probability that the $(n+1)$th word is one of the words that have appeared within the first n words is proportional to its frequency of appearance, and (ii) there is a fixed probability δ that the $(n + 1)$th word will be a new one—a word that has not appeared within the first n words. Assuming that the text is generated accordingly to the Simon model, its frequency-rank dependence will asymptotically follow Zipf's law (1) with exponent (Simon 1955)

$$\alpha = 1 - \delta. \tag{7}$$

Even though the Simon model is not able to reproduce the subsequent part of a text given its preceding part [due to its stochastic origin and the existence of a number of words with the same frequencies (3)], it was shown that its mechanism plausibly describes real writing processes (Holovatch and Palchykov 2007). However, to have a deeper understanding of the language features, one has to go beyond Zipf's law that omits the relationships between interacting words. Below we describe how the structure of these relationships may be studied using the theory of complex networks (Bornholdt and Schuster 2003).

Language Networks

The first step in applying complex network tools to investigate quantitative properties of human language is to represent that language as a network or a graph. Within such an interpretation key language units are considered as the nodes of the network and the links reflect relations between them. Depending on the purpose of investigation, different language units may be used as the network nodes: phonemes, words, concepts or sentences being amongst the possibilities. Then, focusing on one type of node, different networks of language may be reconstructed for various interpretations of the links that connect the nodes, e.g. semantic or syntactic relationships between words as described in the Introduction.

Network Representations

In our investigation we analyse the properties of language networks whose nodes correspond to unique words of a given text. Let us connect a couple of nodes if the corresponding words co-occur at least within a single sentence. We will refer to this representation as the L-space of human language (Fig. 2a). The links in

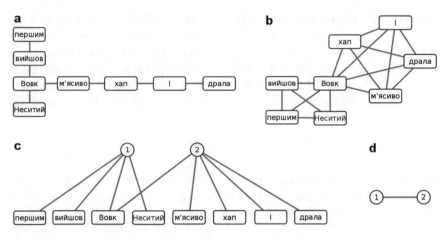

Fig. 2 Networked representation of the corpus of two sentences that share a single common word "вовк" (wolf). Panel (**a**): *L*-space of human language. The co-occurring word-nodes are connected by a link. Panel (**b**): *P*-space. In this space the links connect all word-nodes that belong to the same sentence. Hence, each sentence appears in the network as a complete subgraph (clique). Panel (**c**): *B*-space. This space is represented by a bipartite network that contains two types of nodes: word-nodes and sentence-nodes. The links connect the nodes of different types only: sentences and the words that have been found within. Panel (**d**): *C*-space. Here the nodes represent different sentences and the two nodes are connected if they share at least a single word

L-space as well as in the other spaces that will be described below[3] are unweighted. Alternatively, the links may connect not only the nearest neighbours, bur rather all words that are located on a specific distance from each other. In order to take into account this issue, we introduce a radius of interactions R: for $R = 1$ the links connect only co-occurring words (the nearest neighbours), for $R = 2$ the links connect the nearest and the next nearest neighbours etc. Radius R may be assigned any value in the range $R \in [1, R_{max}]$, where $R_{max} + 1$ is the size of (the number of words within) the longest sentence. For the value $R = 1$ the resulting network reduces to the *L*-space of human language. If $R = R_{max}$ then the links connect all words that belong to the same sentence and the corresponding network representation will be referred to as the *P*-space of human language (Fig. 2b).

An alternative way to represent a language network is to consider it as a bipartite graph (Diestel 2005). We refer to this as *B*-space, see Fig. 2c. Such a representation consists of two different types of nodes: the nodes of one type represent sentences and the nodes of the other type represent words. The link may connect the nodes of different types only and reflect the appearance of given word within a specified sentence. Bipartite networks provide a simple way to obtain two corresponding

---

[3]For different network representations (different spaces) we use the nomenclature originally introduced in the context of transportation networks (Sienkiewicz and Hołyst 2005; von Ferber et al. 2007, 2009).

unipartite networks by performing one-mode projections. One of the resulting networks reproduces the P-space of human language. The other one will be referred to as the language C-space (Fig. 2d). In C-space the nodes represent sentences and the link between two nodes exists if the corresponding sentences have at least one word in common.

The original part of our investigation will be focussed on the networks of language whose nodes represent unique words. Varying the values of R, will allow to perform a comparison between the properties of different networks, whose marginal realizations represent L- and P-spaces of language.

Basic Network Characteristics

To investigate the properties of the language networks we will analyse a number of standard network characteristics: the distribution $P(k)$ of node degrees, average node degree $\langle k \rangle$, average clustering coefficient $\langle C \rangle$ and the average value of the shortest path length $\langle l \rangle$ between the nodes of the network that are briefly described below.

The degree k_i of a node i is defined as the number of links that connect that node with the other nodes of the network. The set of all node degrees $\{k_i\}$ may be characterized by their distribution $P(k)$: the probability that a randomly chosen node has k connections. Modifying the normalization condition one may consider $P(k)$ as the number of nodes whose degree equals to k. The distribution of node degrees contains all the necessary information about fluctuations of the node degree around its average value $\langle k \rangle$ that is defined as

$$\langle k \rangle = \frac{1}{V} \sum_{i=1}^{V} k_i. \tag{8}$$

Here i runs over all V nodes of the network. Both the average degree $\langle k \rangle$ and the shape of the degree distribution $P(k)$ are key characteristics of the global network topology and its local fluctuations.

Networks with identical sequences of degrees may vary significantly due to the variations of the local connectivity patterns or correlations. One kind of correlation, the local grouping of network nodes, may be characterized by a clustering coefficient. The clustering coefficient C_i of node i is defined as the probability that two randomly selected neighbours are connected with each other:

$$C_i = \frac{2m_i}{k_i(k_i - 1)}, \quad k_i > 1, \tag{9}$$

where m_i is the number of links that interconnect the nearest neighbours of node i. The average value of clustering coefficient

$$\langle C \rangle = \frac{1}{V} \sum_{i=1}^{V} C_i \tag{10}$$

characterizes the local grouping in the entire network and may be used to compare the particular network with a random graph that lacks such correlations.

The shortest path length l_{ij} between two nodes i and j is defined as the minimal number of links that should be passed in order to reach node j starting at i. The average shortest path length

$$\langle l \rangle = \frac{2}{V(V-1)} \sum_{i>j} l_{ij} \tag{11}$$

is one of the characteristics of the network. Another characteristics of the network is the maximal value of the shortest path length $l_{max} = \max(\{l_{ij}\})$.

Ukrainian Language Networks

Having introduced the main network characteristics, let us investigate the properties of the language networks of the selected fables. These characteristics for three values of R: $R = 1, 2$ and R_{max} are summarized in Table 2, and will be discussed in detail below.

The distributions of the node degrees $P(k)$ for *Mykyta the Fox* in *L*- and *P*-spaces, which give the number of nodes with degree k, are shown in Fig. 3a. Besides the peak around $k \sim 10$, observed in *P*-space, the tails of the distributions follow a straight line on a double logarithmic scale, and the functional dependence of the node degree distributions may therefore be described by the power law function. Similar dependencies describe degree distributions of *Abu-Kasym's slippers* and the combination of the two fables. The exponent of the power law decay function (2) fluctuates around $\gamma = 1.9$ to 2.0, but the sizes of the investigated networks do not allow one to make more precise estimations. In order to justify the power law behaviour of the degree distribution, we additionally consider the cumulative node degree distribution:

$$P_{cum}(k) = \sum_{k'=k}^{k_{max}} P(k'). \tag{12}$$

Table 2 The basic quantitative characteristics of the investigated language networks for several values of R. The upper part of the table corresponds to the fable *Abu-Kasym's slippers*, the middle part corresponds to *Mykyta the Fox* and the bottom part represents the combined text of the two fables

R	V	M	$\langle k \rangle$	k_{max}	γ	γ_{cum}	$\langle C \rangle$	$\langle C \rangle / C_r$	$\langle l \rangle$	l_{max}
1	2392	6273	5.24	228	1.9	1.2	0.172	78	3.43	11
2	2392	11,475	9.59	391	2.0	1.2	0.567	141	2.90	7
R_{max}	2392	48,603	40.64	1134	1.9	1.4	0.841	50	2.22	4
1	3563	11,102	6.23	419	1.9	1.1	0.214	122	3.30	11
2	3563	20,063	11.26	665	1.8	1.2	0.588	186	2.85	7
R_{max}	3563	65,997	37.05	1526	1.9	1.3	0.822	79	2.27	5
1	4823	16,580	6.88	537	1.9	1.1	0.243	170	3.24	11
2	4823	29,916	12.41	868	1.8	1.2	0.585	227	2.83	7
R_{max}	4823	107,750	44.68	2185	2.0	1.3	0.818	88	2.50	5

The table contains the number of nodes V; the number of links M; average and maximal node degrees $\langle k \rangle$ and k_{max} respectively; the exponents γ and γ_{cum} of the power-law fit to the degree distribution (2) and cumulative degree distribution (12) respectively; the average clustering coefficient $\langle C \rangle$ from Eq. (10) and its counterpart C_r of the corresponding random graph; and the average and maximal shortest path lengths between nodes $\langle l \rangle$ and l_{max}

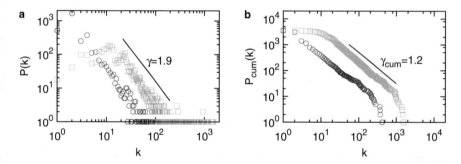

Fig. 3 *Mykyta the Fox*: node degree distributions (panel (**a**)) and cumulative node degree distributions (panel (**b**)) for $R = 1$ (*brown circles*) and $R = R_{max}$ (*grey squares*). The *solid lines* are shown as the guides to the eye and represent the power law decay functions (2) in panel (**a**) and (12) in panel (**b**) with exponents $\gamma = 1.9$ and $\gamma_{cum} = 1.2$, correspondingly

The corresponding dependencies for *Mykyta the Fox* are shown in Fig. 3b. The function $P_{cum}(k)$ is smoother than $P(k)$ and allows one to make a direct conclusion on the power-law behaviour of the degree distribution function.

A similar analysis for the English language has been performed for the British National Corpus[4] with $V \sim 10^7$. The analysis of this corpus (Ferrer i Cancho and Solé 2001b) demonstrates that the corresponding language network is scale-

---

[4]The British National Corpus is a collection of samples of written and spoken language from a wide range of sources, designed to represent a wide cross-section of British English from the late twentieth century, http://www.natcorp.ox.ac.uk/.

free, with power-law decay of the degree distribution $P(k)$ characterized by two distinct regimes with the exponent $\gamma = 1.5$ for $k \leq 2000$ and $\gamma = 2.7$ for $k \geq 2000$. We cannot, of course, claim that the values of the exponents which we have obtained for the two fables persist for the entire corpus of the Ukrainian language. Nor can we exclude the possibility of a crossover in the entire corpus of the type observed in Ferrer i Cancho and Solé (2001b). However, the results of this pilot study demonstrate that the network of the Ukrainian language used, at least for the two fables analysed, is characterized by a scale-free topology.

Besides exhibiting scale-free topologies, many real networks tend to be small worlds (Albert and Barabási 2002; Watts 1999). A network is considered to be a small world if its average shortest path length $\langle l \rangle$ increases with the number of nodes V slower that any power-law function (Dorogovtsev and Mendes 2003). Note for comparison that a regular d-dimensional lattice has $\langle l \rangle \sim V^{1/d}$. Small worlds are extremely compact; an arbitrary pair of nodes is separated just by a few links. The notion is known in sociology (where it originated), where it has been shown that two randomly chosen members of society are separated by an average of six intermediate relationships (Milgram 1967). Table 1 shows that for $R = 1$ the maximal shortest path length for all three networks is $l_{max} = 11$ and the average shortest path length is $\langle l \rangle \sim 3$. The average $\langle l \rangle$ and the maximal l_{max} decreases with R, reaching about 4 or 5 for l_{max} and just above 2 for $\langle l \rangle$ in P-space. Such behaviour is quite natural, since the number of links may only increase (or remain unchanged) with increasing R without affecting the number of nodes V. This extreme compactness suggests that the Ukrainian language networks used in the fables are characterized by the small world effect, even though a strict conclusion would require a study of size dependent evolution. For comparison, the average shortest path length $\langle l \rangle$ for the above-mentioned English resource is $\langle l \rangle = 2.63$ (Ferrer i Cancho and Solé 2001b).

Connected triangles of nodes are typical signs for the presence of correlations in networks. Defined in Eq. (10), the average clustering coefficient $\langle C \rangle$ is expected to characterize this type of correlation. The clustering coefficient for a complete graph is $\langle C \rangle = 1$ and it is $\langle C \rangle = 0$ for a tree-like network. To characterize the level of these local correlations, the clustering coefficient of a network is usually compared to the one of a random graph with the same number of nodes V and links M, for which

$$C_r = \frac{2M}{V^2}. \tag{13}$$

Table 2 gives average values of the clustering coefficients $\langle C \rangle$ and their ratios to the those for the corresponding random graphs C_r. Since the observed clustering coefficients $\langle C \rangle$ vastly exceed their random counterparts C_r, the networks under consideration are well correlated structures. As expected, these correlations become stronger with the radius of interaction R, see Table 2.

Finally, let us investigate the influence of the node degree k on the distance that separates the node from the rest of the network and on the connectivity patterns of its neighbourhood. The first of these (distance) is quantified by the average

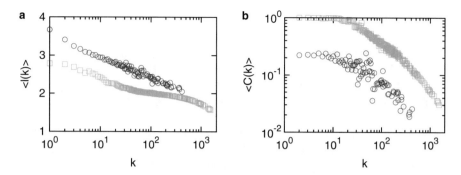

Fig. 4 *Mykyta the Fox*: average shortest path length $\langle l(k) \rangle$ that separates a node of degree k from the other nodes of the network as a function of k (panel (**a**)), and the dependence of the average clustering coefficient $\langle C(k) \rangle$ of a node with k neighbours on k (panel (**b**)). The L- ($R = 1$) and P-spaces ($R = R_{max}$) are represented by *dark circles* and *grey squares*, respectively

shortest path length $\langle l(k) \rangle$ from a node of degree k to an arbitrary chosen reachable network node. The connectivity patterns of the neighboring nodes are described by the average clustering coefficient $\langle C(k) \rangle$ of the nodes of degree k. Figure 4a shows that the average distance $\langle l(k) \rangle$ from a node of degree k to a randomly selected other node of the network monotonically decreases with k. This reflects the observation that the higher the degree of the node, the less distance separates the node from the rest of the network. This short-distance scenario for the hub nodes is complemented by the second scenario: the more neighbours the node has, the less connected these neighbours are amongst each other, see Fig. 4b. A high level of clustering coefficient is observed only for the nodes with small degree, and it decreases rapidly for higher degree nodes.

It is interesting to compare the results of our analysis with studies of collections of texts written in Portuguese and English, whose sizes ranged between 169 and 276,425 words (Caldeira et al. 2006). The representation of texts used in Caldeira et al. (2006) corresponds to *P*-space in our classification. However, unlike our investigation, a network of concepts has also been considered in Caldeira et al. (2006) and could form the basis for a future study of fables. The quantitative characteristics of the corresponding network are $\gamma = 1.6 \pm 0.2$, $\langle l \rangle = 2.0 \pm 0.1$, $l_{max} = 4 \pm 1$, $\langle C \rangle = 0.83 \pm 0.03$ and are in a good agreement with our results ($\gamma = 2.0$, $\langle l \rangle = 2.25$, $l_{max} = 5$, $\langle C \rangle = 0.818$ for the combination of two fables). Such agreement is interesting, not only because the investigated networks correspond to different languages (English, Portuguese in Caldeira et al. (2006) and Ukrainian in our case), but mainly because it shows that the restriction to the specific subset of words (words-concepts) does not cause significant changes in the above-considered features of a language network.

Conclusions

In this article, a quantitative analysis of the word distribution in two fables written in Ukrainian by Ivan Franko (*Mykyta the Fox* and *Abu-Kasym's slippers*) has been performed. Our investigation consists of two distinct parts: analysis of the frequency-rank dependence (Zipf's law) and a deeper analysis of the structure of language using the tools developed within complex network science. A main purpose of analyzing the frequency-rank dependencies is to verify whether the texts under investigation are large enough to exhibit the expected statistical features. Having confirmed the validity of Zipf's law for the rank values in the range $r = 20$–3000, we have justified that these texts may be used to investigate deeper features of the Ukrainian language used in those fables.

The results of our investigation confirm that the network of the Ukrainian language used is a highly correlated, small world that is characterized by a scale-free topology. This is quite expected, since similar results have been formerly obtained for various other languages (Ferrer i Cancho and Solé 2001b; Caldeira et al. 2006; Zhou et al. 2008). The small world effect highlights an extremely high level of compactness of such networks: despite the large size of the vocabulary used, a pair of randomly selected words is separated on average by only three steps. The high values for clustering coefficients observed in our analysis reflect the high level of correlations in the network structures. The empirical results obtained here may be used for a theoretical description of the evolution of language, which may be based on evolutionary game theory (Nowak and Krakauer 1999).

Some attempts have been made to explain features of language networks using the preferential attachment scenario (Albert et al. 1999); one attempts to consider such networks as the results of a growth process, wherein new words that join the network tend to be connected to hubs with higher probability than to low degree nodes (Ferrer i Cancho and Solé 2001b; Caldeira et al. 2006; Dorogovtsev and Mendes 2001). It is worth noting that, following such an approach, the emergence of syntax is a consequence of the evolution of language (Ferrer i Cancho et al. 2005; Solé 2005).

Our analysis of language networks in different spaces shows that basic features, such as small world effect or scale-free topology, are space independent and highlight the properties of the language itself rather than of a particular representation. This is strongly confirmed by our comparison of distinct networks in P-space, which shows that the restriction towards some categories of words does not cause significant changes in the features of these networks.

In this chapter we report on an attempt to analyze some properties of the Ukrainian language based on two narratives. On the one hand, indeed one has to take a larger database to study the entire language. On the other hand, as we show in section "Word Appearance Statistics", the universal properties of word distribution already hold for our database. It gives hope that some universalities of language can be checked using this database too. Moreover, in similar studies texts of similar word numbers have been used, see e.g. Caldeira et al. (2006). An interesting question for further analysis might be to search the non-universal, specific properties of the language used in a given narrative or rather in a given group of narratives of a similar type. Could one identify (or even categorise) genres through such properties? A question of interest for the main readership (and authors) of this book probably would be whether quantitative features of the language of mythological narratives make them different from other texts? We think that analysis performed from such perspectives may lead to better understandings of what is still hidden in old narratives. In other words, this is a first step of a new programme and there is much more to do.

Acknowledgements It is our pleasure to thank the Editors of this book Ralph Kenna, Máirín MacCarron, and Pádraig MacCarron for their invitation to contribute and for their help and discussions during preparation of the manuscript. Yu.H. acknowledges useful discussions with Bernat Corominas-Murtra. This work was supported in part by the 7th FP, IRSES projects No. 295302 Statistical Physics in Diverse Realizations (SPIDER), No. 612707 Dynamics of and in Complex Systems (DIONICOS), by the COST Action TD1210 Analyzing the dynamics of information and knowledge landscapes (KNOWSCAPE) and by SNSF project No. 147609 Crowdsourced conceptualization of complex scientific knowledge and discovery of discoveries.

References

Albert, R., & Barabási, A. L. (2002). Statistical mechanics of complex networks. *Reviews of Modern Physics, 74*, 47–97. doi:10.1103/RevModPhys.74.47.

Albert, R., Jeong, H., & Barabási, A.-L. (1999). Diameter of the World-Wide Web. *Nature, 401*, 130–131. doi:10.1038/43601.

Barceló-Coblijn, L., Corominas-Murtra, B., & Gomila, A. (2012). Syntactic trees and small-world networks: Syntactic development as a dynamical process. *Adaptive Behavior, 20*(6), 427. doi:10.1177/1059712312455439.

Borge-Holthoefer, J., & Arenas, A. (2010). Semantic networks: Structure and dynamics. *Entropy, 12*, 1264–1302. doi:10.3390/e12051264.

Bornholdt, S., & Schuster, H. (Eds.). (2003). *Handbooks of graphs and networks*. Weinheim: Wiley.

Caldeira, S. M. G., Petit Lobão, T. C., Andrade, R. F. S., Neme, A., & Miranda, J. G. V. (2006). The network of concepts in written texts. *European Physical Journal B: Condensed Matter and Complex Systems, 49*, 523–529. doi:10.1140/epjb/e2006-00091-3.

Condon, E. U. (1928). Statistics of vocabulary. *Science, 67*, 300. doi:10.1126/science.67.1733.300.

Corominas-Murtra, B., Hanel, R., & Thurner, S. (2015). Understanding scaling through history-dependent processes with collapsing sample space. *Proceedings of the National Academy of Sciences of the United States of America, 112*, 5348–5353. doi:10.1073/pnas.1420946112.

Corominas Murtra, B., Valverde, S., & Solé, R. V. (2007). Emergence of scale-free syntax networks. Preprint. arXiv:0709.4344.

de Jesus Holanda, A., Torres Pisa, I., Kinouchi, O., Souto Martinez, A., & Seron Ruiz, E. E. (2004). Thesaurus as a complex network. *Physica A, 344*, 530–536. doi:10.1016/j.physa.2004.06.025.

Diestel, R. (2005). *Graph theory (Graduate texts in mathematics)*. Heidelberg: Springer.

Dorogovtsev, S. N., & Mendes, J. F. F. (2001). Language as an evolving word web. *Proceedings of the Royal Society B, 268*, 2603–2606. doi:10.1098/rspb.2001.1824.

Dorogovtsev, S. N., & Mendes, J. F. F. (2003). *Evolution of networks*. Oxford: Oxford University Press.

Estoup, J. B. (1916). *Gammes stenographiques*. Paris: Institut Stenographique de France.

Ferrer i Cancho, R. (2005). The variation of Zipf's law in human language. *European Physical Journal B: Condensed Matter and Complex Systems, 44*, 249–257. doi:10.1140/epjb/e2005-00121-8.

Ferrer i Cancho, R., Riordan, O., & Bollobás, B. (2005). The consequences of Zipf's law for syntax and symbolic reference. *Proceedings of the Royal Society B, 272*, 561–565. doi:10.1098/rspb.2004.2957.

Ferrer i Cancho, R., & Solé, R.V. (2001a). Two regimes in the frequency of words and the origin of complex lexicons: Zipf's law revisited. *Journal of Quantitative Linguistics, 8*, 165–173. doi:1076/jqul.8.3.165.4101.

Ferrer i Cancho, R., & Solé, R. V. (2001b). The small world of human language. *Proceedings of The Royal Society of London. Series B, Biological Sciences, 268*, 2261–2265. doi:10.1098/rspb.2001.1800.

Ferrer i Cancho, R., Solé, R. V., & Köhler, R. (2004). Patterns in syntactic dependency networks. *Physical Review E, 69*, 051915. doi:10.1103/PhysRevE.69.051915.

Holovatch, Yu., & Palchykov, V. (2007). Mykyta the Fox and networks of language. *Journal of Physical Studies, 11*, 22–33 (in Ukrainian).

Kanter, I., & Kessler, D. A. (1995). Markov processes: Linguistics and Zipf's law. *Physical Review Letters, 74*, 4559–4562. doi:10.1103/PhysRevLett.74.4559.

Li, W. (1992). Random texts exhibit Zipf's-law-like word frequency distribution. *IEEE Transactions on Information Theory, 38*, 1842–1845. doi:10.1109/18.165464.

Masucci, A. P., & Rodgers, G. J. (2006). Network properties of written human language. *Physical Review E, 74*, 026102. doi:10.1103/PhysRevE.74.026102.

Milgram, S. (1967). The small-world problem. *Psychology Today, 2*, 61–67.

Montemuro, M. A. (2001). Beyond the Zipf-Mandelbrot law in quantitative linguistics. *Physica A, 300*, 567–578. doi:10.1016/S0378-4371(01)00355-7.

Motter, A. E., de Moura, A. P. S., Lai, Y.-C., & Dasgupta, P. (2002). Topology of the conceptual network of language. *Physical Review E, 65*, 065102(R). doi:10.1103/PhysRevE.65.065102.

Newman, M. E. J. (2010). *Networks: An introduction*. Oxford: Oxford University Press.

Newman, M. E. J. (2012). Communities, modules and large-scale structure in networks. *Nature Physics, 8*, 25–31. doi:10.1038/nphys2162.

Nowak, M. A., & Krakauer, D. C. (1999). The evolution of language. *Proceedings of the National Academy of Sciences of the United States of America, 96*, 8028–8033. doi:10.1073/pnas.96.14.8028.

Petersen, A. M., Tenenbaum, J. N., Havlin, S., Stanley, H. E., & Perc, M. (2012). Languages cool as they expand: Allometric scaling and the decreasing need for new words. *Scientific Reports, 2*, 943. doi:10.1038/srep00943.

Sienkiewicz, J., & Hołyst, J. A. (2005). Statistical analysis of 22 public transport networks in Poland. *Physical Review E, 72*, 046127. doi:10.1103/PhysRevE.72.046127.

Sigman, M., & Cecchi, G. A. (2002). Global organization of the Wordnet lexicon. *Proceedings of the National Academy of Sciences of the United States of America, 99*, 1742. doi:10.1073/pnas.022341799.

Simon, H. A. (1955). On a class of skew distribution functions. *Biometrica, 44*, 425–440. doi:10.1093/biomet/42.3-4.425.

Solé, R. (2005). Syntax for free? *Nature, 434*, 289. doi:10.1038/434289a.

Solé, R. V., Corominas-Murtra, B., Valverde, S., & Steels, L. (2010). Language networks: Their structure, function, and evolution. *Complexity, 15*(6), 20. doi:10.1002/cplx.20305.

Solé, R. V., & Seoane, L. F. (2014). Ambiguity in language networks. *The Linguistic Review, 32*(1), 5–35. doi:10.1515/tlr-2014-0014.

Thurner, S., Hanel, R., Liu, B., & Corominas-Murtra, B. (2015). Understanding Zipf's law of word frequencies through sample-space collapse in sentence formation. *Journal of the Royal Society Interface, 12*, 20150330. doi:10.1098/rsif.2015.0330.

von Ferber, C., Holovatch, T., Holovatch, Yu., & Palchykov, V. (2007). Network harness: Metropolis public transport. *Physica A, 380*, 585–591. doi:10.1016/j.physa.2007.02.101.

von Ferber, C., Holovatch, T., Holovatch, Yu., & Palchykov, V. (2009). Public transport networks: Empirical analysis and modeling. *European Physical Journal B, 68,* 261–275. doi:10.1140/epjb/e2009-00090-x.

Watts, D. J. (1999). *Small words.* Princeton, NJ: Princeton University Press.

Zhou, S., Hu, G., Zhang, Z., & Guan, J. (2008). An empirical study of Chinese language networks. *Physica A, 387,* 3039–3047. doi:10.1016/j.physa.2008.01.024.

Zipf, G. K. (1935). *The psycho-biology of language.* Boston: Houghton-Mifflin.

Zipf, G. K. (1949). *Human behaviour and the principle of least effort. An introduction to human ecology* (1st ed.). Cambridge: Addison-Wesley (Hafner reprint, New York, 1972).

Analysing and Restoring the Chronology of the Irish Annals

Daniel Mc Carthy

Abstract Substantial annalistic chronicles of Irish affairs exist in a number of medieval versions, but they exhibit considerable variation both in the sequences of events and the chronological apparatus used to link each year to the Julian calendar. Of these, the Anno Domini years of the *Annals of Ulster* have been principally relied upon by historians. However, these are demonstrably incorrect from the seventh to the eleventh centuries. Moreover, its remaining chronological data of ferials and lunar epacts at the kalends of January, that is, the day of the week and the age of the moon on 1 January, are almost all interpolations by a later scribe. On the other hand, the *Annals of Tigernach* and the *Chronicum Scotorum* have only kalends and ferials marking the commencement of each year from the Incarnation up until the mid-seventh century. Because these kalends and ferials are susceptible to scribal miscopying they were dismissed by historians and textual scholars as "hopelessly confused". However, analysis of the 28 year cycle of the ferials reveals that they possess a powerful error-correction property. Exploitation of this property has enabled the restoration of all the missing kalends and erroneous ferials of the *Annals of Tigernach* and *Chronicum Scotorum*, as well as of the closely related *Annals of Roscrea*, known collectively as the Clonmacnoise group. Using computer table structures, the kalends and ferials and events of these three have been synchronized with the Anno Domini years over the range AD 1–1178, and this tabulation, with cross-references to the other Irish medieval annals, has been made available online at www.irish-annals.cs.tcd.ie. In this chapter the process of analysis, correction, and synchronization is illustrated, taking the year of the death of St Patrick as an example.

Introduction: The Irish Annals

Ireland has a remarkable legacy in terms of both the quantity and quality of its medieval chronicles. These are preserved in medieval manuscripts and they contain sequences of textual entries distributed across time as systematically represented by

D. Mc Carthy (✉)
Department of Computer Science, Trinity College Dublin, Dublin, Ireland
e-mail: Dan.McCarthy@cs.tcd.ie

© Springer International Publishing Switzerland 2017
R. Kenna et al. (eds.), *Maths Meets Myths: Quantitative Approaches to Ancient Narratives*, Understanding Complex Systems,
DOI 10.1007/978-3-319-39445-9_10

their chronological apparatus. Collectively they extend from "Creation" to the year 1616 AD. The words "chronicle" and "manuscript" refer to two different things; the manuscript is the physical tome which contains a copy of the chronicle. For example, four different recensions (text revisions) of the narrative chronicle *Lebor Gabála* (the *Book of Invasions*) have come down to us in more than a dozen medieval manuscripts, such as, for example, in the *Book of Leinster* (dating from circa AD 1150), and the *Book of Ballymote* (from AD 1391). The chronological apparatus of *Lebor Gabála* represents time using the number of years of the reigns of a long, standardised series of rulers, known to the medieval chronologists as the *Réim Rígraide* and in English as the 'regnal canon'. Some chronicles explicitly represent almost every year and these are known as 'annals', from the Latin *annales*. This annalistic genre was the most important for chronicles compiled in Ireland. Ten substantial collections of Irish annals have come down to us, namely: *Annals of Tigernach, Chronicum Scotorum, Annals of Roscrea, Annals of Ulster, Annals of Inisfallen, Annals of Boyle, Annals of Connacht, Annals of Loch Cé, Mageoghagan's Book* and *Annals of the Four Masters*.

The manuscripts of these annals date from the eleventh to the seventeenth century. These were clearly copied from earlier manuscripts, and in the process of transcription they underwent some re-compilation and supplementary composition. The general consensus is that one of the earliest is an annalistic chronicle which was kept in the monastery at Iona up to the mid-eighth century. In circa AD 740 this, or a copy, was relocated to Ireland where it was continued by ecclesiastic scholars in the midlands up until the earlier thirteenth century. This ecclesiastical chronicle was then further continued by secular Gaelic scholars up to the end of the sixteenth century, and finally by synthesising compilers in the early seventeenth century (see, e.g., Mac Niocaill 1975, pp. 18–37; Mc Carthy 2008, pp. 355–357).

The entries of these annals are written variously in three languages—Latin, Irish, or English, and these entries may refer either to earlier or contemporaneous events. In the latter case the entry may be referred to as a *record*. If an entry is believed to be a later addition to the chronicle, it is called an *interpolation*. Besides the annals, there are broadly two other genres of chronicle—narrative and poetic. Table 1 lists 15 Irish chronicles. The first 10 are substantially annalistic, as indeed is implied by the usage of the word "Annals" in most of their published titles. The next three in the list are narrative, and the final two are poetic.

The annalistic form of chronicle arrived in Ireland with Christianity and the Irish maintained it down to the seventeenth century. It is, indeed, the most common format used in Irish chronicles. Moreover, no other European literate class compiled this genre to the extent of the Irish. Thus these annals represent a substantial and important literary and historical resource, preserving not only elements of culture from the fifth century, but also reflecting how Irish scholarship adapted to Christianity, and providing a chronological framework for all of Ireland's medieval history.

The annals from before the earlier fifth century mainly reflect their Mediterranean Christian origin, so that their content is not intrinsically Irish. Subsequent annals shift the focus first onto the affairs of the east, then the north of Ireland as

Table 1 Fifteen Irish chronicles, their sigla, genres, chronological apparatus, and annalistic groups. The published editions of all of these chronicles are listed in the Bibliography

Chronicle	Siglum	Genre	Chronological apparatus	Annalistic group
Annals of Tigernach	AT	Annalistic	Kalend	Clonmacnoise
Chronicum Scotorum	CS	Annalistic	Kalend	Clonmacnoise
Annals of Roscrea	AR	Annalistic	Kalend	Clonmacnoise
Annals of Ulster	AU	Annalistic	Kalend	Cuana
Annals of Inisfallen	AI	Annalistic	Kalend	Cuana
Annals of Boyle	AB	Annalistic	Kalend	Cuana
Annals of Connacht	CT	Annalistic	Kalend	Connacht
Annals of Loch Cé	LC	Annalistic	Kalend	Connacht
Conell Mageoghagan's Book	MB	Annalistic	Regnal-canon	Regnal-canon
Annals of The Four Masters	FM	Annalistic	Regnal-canon	Regnal-canon
Cogadh Gaedhel re Gallaibh	CG	Narrative	Regnal-canon	–
Lebor Gabála	LG	Narrative	Regnal-canon	–
Foras Feasa ar Eirinn	FF	Narrative	Regnal-canon	–
Gilla Cóemháin's poems	GC	Poetic	Regnal-canon	–
Flann Mainistreach's poems	FL	Poetic	–	–

well as Scotland. In the mid-eighth century the Scottish element quite suddenly diminishes and the focus thereafter remains broadly Irish. The emphasis first turns to Clonmacnoise in the Irish midlands, and then to Connacht in the west from the thirteenth century onwards. Over the ninth and tenth centuries, ecclesiastical affairs dominate the annalistic purview (Etchingham 2002). However, from the second half of the tenth century, a significant increase in the inclusion of secular events takes place. Thus, from the fifth to the seventeenth century, the annals supply cryptic accounts of thousands of events relating to Ireland and her people. Typically these record events such as obits (obituaries) of ecclesiastical and secular leaders, battles, catastrophes such as plagues and famines, astronomical phenomena such as eclipses and comets, and meteorological extremes of rain, cold, and drought. Up until the fifth century the language employed is mostly Latin, but thereafter personal names and place names are generally registered in Irish. This is followed by an inexorable shift to the vernacular in subsequent centuries, so that by the tenth century about half of the *Annals of Ulster*—the text preserving most of the oldest orthography—is in Irish (Dumville 1982, pp. 329–332).

Of the 10 annalistic chronicles listed in Table 1, eight identify each year by either a 'K', 'Kl', or 'Kł', representing 'Kalendae Ianuarii', the first day of January, and thus are designated as belonging to the *Kalend tradition*. These eight may be further divided into three groups. The *Clonmacnoise group* (AT, CS, AR) is so named because from the mid-eighth century they focus on the affairs of the monastery of that name situated in the Irish midlands, and on its environs. Textually, these three share many common features and close semantic relationships between their common entries, as well as virtually identical chronologies and chronological apparatus, which, up until the mid-seventh century, comprise simply the kalend followed by the ferial (week day) of 1 January. The *Cuana Group* (AU, AI, AB) takes its name from the now lost "Liber Cuanach", compiled by Cuána Ua Lothcháin †1024. Their chronological apparatus includes 19-year lunar epacts (the age of the moon on 1 January). The *Connacht Group* (CT, LC) has extensive coverage of the affairs of that province, and its two members share a common chronological apparatus which includes numerous chronological criteria. The remaining two annalistic chronicles, *Mageogagan's Book* and the *Annals of the Four Masters*, are the only ones which employ a regnal chronology, following the kings of Ireland from the Fir Bolg, a race believed by the medieval historians to have inhabited Ireland in ancient times, up to the death in 1022 of Máel Sechnaill mac Domhnaill, king of Meath and high king of Ireland. Thus these two are designated as belonging to the *Regnal-canon tradition*. These various classifications are shown in Table 1.

The medieval system of nomenclature for these chronicles used the word "liber", or its Irish equivalent "leabhar", together with either the name of the place it was compiled, or the name of the perceived compiler. Thus *Leabhar Cluana Mic Nóis* (*Book of Clonmacnoise*), or *Liber Cuanach* (*Cuána's Book*). However, for the annals listed in Table 1, with the exceptions of *Chronicum Scotorum* and *Annales Roscreenses*, the modern titles have no contemporaneous or manuscript basis. The present titles of AT, AU, AI, AB, LC, and CT were suggested by either James Usher or James Ware in the seventeenth century. Regarding FM, while Mícheál Ó Cléirigh, its principal compiler, appears to have used the Latin title "Annales Regni Hiberniae", this was effectively superseded by John Colgan's substitution of *Annales Quatuor Magistrorum* (*Annals of the Four Masters*) in 1645.

Some annalistic titles have been justifiably challenged in modern times. For example, the compilation that we refer to as *Mageoghagan's Book* is a 1627 translation into English by Conell Mageoghagan of what he called an "old Irish booke", which is now lost. It was Ware who referred to Mageoghagan's compilation as the "Annals of Clonmacnoise" on the basis of the number of entries referring to the monastery of Clonmacnoise and its hinterland. This was subsequently repeated by Denis Murphy who employed it for the title of his 1896 edition, albeit with some misgivings. Naturally any reference to Murphy's edition must acknowledge that title, but nevertheless repeated reservations about the title have been expressed. In particular, *Mageoghagan's Book* employs a regnal-canon chronological apparatus rather than the kalend and ferial chronological apparatus characteristic of the Clonmacnoise Group (AT, CS, AR), and, moreover, it is not in fact annalistic until the middle of the seventh century. Therefore, it is a poor representative of that group and Ware's title is misleading, and for these reasons I prefer to refer to it as *Mageoghagan's Book*.

While the annals are a rich source of historical information, there are consider-able difficulties in using them for the earlier Irish Christian period because in most cases their chronological apparatus do not provide an Anno Domini year. In these circumstances the editors of the published editions have endeavoured to make good the deficiency by inserting a marginal editorial AD year, but these can vary widely. A well-known example concerns the year of the death of St Patrick, Ireland's patron saint. In the published edition of the *Annals of Innisfallen*, which are preserved in the earliest annalistic manuscript, the year is given by the editor as AD 496; the *Annals of Ulster* is the version most frequently relied upon by historians, and in Hennessy's nineteenth century edition, three separate entries are given at the years AD 461, 491 and 492, while in the more recent edition of Mac Airt and Mac Niocaill these entries are assigned to the years AD 461, 492 and 493; in the *Annals of the Four Masters*, popularly regarded as the pre-eminent annals, it is given at AD 493; in Hennessy's edition of *Chronicum Scotorum* he assigns the year to AD 489; in Stokes' edition of the *Annals of Tigernach* he cites the 4 years AD 488, 489, 492, and 493. These variants extending from AD 461 to AD 496 have inevitably contributed a great deal of confusion regarding the chronology of the death of St Patrick.

However, by exploiting the aforementioned error-correction property of the 28 year cycle of the ferials of the Clonmacnoise group it is possible to identify all the kalend and ferial errors and so correct any missing or interpolated kalends and to synchronize the chronicle years with the Anno Domini years. In the remainder of this chapter, I illustrate the process used.

Kalends and Ferials

The Romans did not use a forward counting system to reckon the days of the month as we do. Instead they counted backwards from three fixed days of each month: the Kalends, the Nones and the Ides. The Kalends is the first day of the month; The Nones, is the seventh day of March, May, July and October, or the fifth day of the other 8 months. The Ides is the 15th day of March, May, July and October, or the 13th day of the other 8 months. For example, 16th kalends of April means the 16th day counting backwards inclusively from the first of April. Thus 16th kalends April is 17 March in modern terms.

A particular feature of those Irish annals which are contained in manuscripts compiled before the seventeenth century is that they all identify the beginning of each year by the inscription 'K', 'Kl', or 'Kł', representing 'Kalendae Ianuarii' or the first of January. This method of identifying chronicle years is unique to the Irish annals, and was a mystery until in 1985 Dáibhí Ó Cróinín discovered an Easter table in a manuscript preserved in the Biblioteca Antoniana in Padua, Italy, which used the same system (Mc Carthy and Ó Cróinín 1990, pp. 228–229). This manuscript, written in northern Italy in the early years of the tenth century, preserves a full copy of the 84-year Easter table known to have been employed by the early churches in Ireland and Britain up to the eighth century (see Fig. 1). The data in the table, its heading, and the context all suggest that this Easter table is one of the earliest documents used by the early Irish churches. Ó Cróinín's discovery was truly remarkable, for there had been no recorded sighting of this 84-year Easter table since AD 716. In this table each year is identified with 'Kl', representing 'Kalendae Ianuarii', followed by the ferial of that day, just as in the Clonmacnoise group of the Irish Annals.

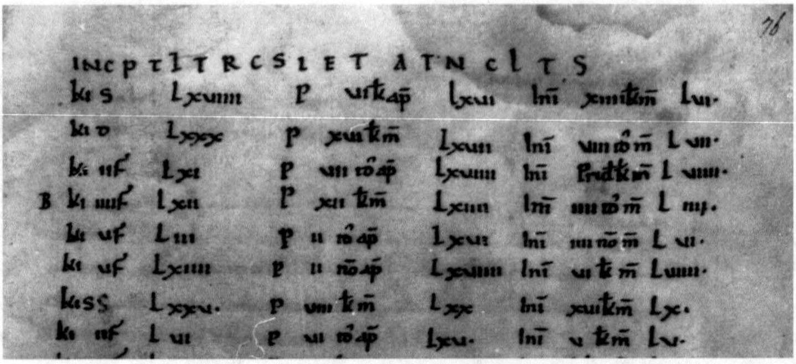

Fig. 1 The heading and first 8 years of the 84-year Easter table preserved in the manuscript, Padua I 27, folios 76r–77v. Each year is identified by 'Kl' followed by the weekday of 1 January. Reproduced by permission of The Librarian, Biblioteca Antoniana, Padua

Figure 1 reproduces the part of Padua Easter table containing the first 8 years. Note that each 'Kl' is followed either by the letter 'S', or the letter 'D', or by a Roman numeral prefixed to 'f'. These notations represent the days of the week: 'S' for 'Sabbatum', or Saturday, 'D' for 'Dominicus', or Sunday, and "iif" to "uif" for the remaining 5 days. Monday is 'ii feria', Tuesday is 'iii feria' and so on with Friday being 'ui feria'. Monday is counted as the second day because Sunday is considered the first. Thus, in the Padua Easter table each year is identified by the day of the week on which 1 January falls. For example, "Kl iif" at the third row identifies a year in which the first of January was a Monday, while the first and second years commenced on a Saturday and on a Sunday respectively. Although all earlier Irish annals mark the beginning of each year by 'K', 'Kl', or 'Kł', only the members of the Clonmacnoise group employ ferials to identify each year up until the mid-seventh century.

To give an example of how such information appears in an annalistic manuscript, part of a folio from Dubhaltach Mac Fhirbhisigh's *Chronicum Scotorum* is reproduced in the left panel of Fig. 2. The right panel contains Hennessy's nineteenth-century English translation. Mac Fhirbhisigh was a member of a leading north Connacht family and one of the last traditionally trained Gaelic scholars. During 1665 and 1666, he was employed by James Ware to provide translations into English of Irish texts. The excerpt commences with an entry registering the birth of St Patrick, and note how his name is celebrated with the enlarged and decorated letter "P". Note also the series of 11 'Kł', each accompanied by the ferial number delimited by two periods, and the first two "Kł" are followed by the word "Enair", explicitly indicating that it is the kalends of January. 'Kł.i' identifies a year in which 1 January fell on a Sunday, and 'K.uii.' a year in which 1 January fell on a Saturday. Thus this annalistic chronological apparatus corresponds closely with that used in the Padua Easter table.

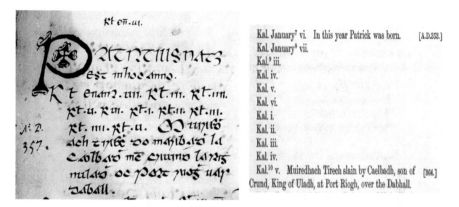

Fig. 2 The *left panel* shows part of folio 164r in the manuscript Trinity College 1292 containing *Chronicum Scotorum* written by Dubhaltach Mac Fhirbhisigh in circa 1640. The kalends and ferials commence with 'Kł En*air*..ui. Patrtius natus est in hoc anno' (Kal. January vi. In this year Patrick was born). Reproduced by permission of the Board of Trinity College, Dublin. The *right panel* shows William Hennessy's 1866 English translation of the excerpt

This is how Hennessy describes this and subsequent text (Hennessy 1866, p. xlii):

> The first entry in the Christian period is the record of the birth of St. Patrick, which is preceded by the criteria 'Kł. Enair, ui,' 'Kal. of January 6,' implying that the kalends, or first, of January occurred on the 6th day of the week, or Friday. The succession of years is then regularly indicated by the repetition of the characters Kł., or K. for 'kalends,' accompanied, with some exceptions, as far as the year 641, by the feriæ, or days of the week on which the first of January fell in each year. Subsequently to the date 641, the feriæ are no longer noted, every year being simply marked by the sign 'Kł'.

In the left panel of Fig. 3, we reproduce that part of Hennessy's 1866 transcription of *Chronicum Scotorum* which registers the death of St Patrick and in the right panel we again provide Hennessy's translation. In the introduction to his edition Hennessy turned his attention to the question of the chronology of these annals, and wrote (p. xlii):

> The chronology of the following chronicle is in a state of much confusion, notwithstanding the apparent regard for a regular system, indicated by the array of ferial numbers with which the Christian period of the work begins. The feriæ, however, do not run on in consecutive order, owing probably, in large measure, to mistakes committed in the course of successive transcriptions of the original. Much of the confusion created in this respect is traceable to the ease with which the numeral u, as written in old MSS., may be confounded with ii.

Thus scholars have long been aware of problems with the annalistic chronology and it is only in recent years that progress has been made in overcoming them. Remarkably, even though both *Annals of Tigernach* and the *Chronicum Scotorum* have some substantial gaps in their records, largely due to the loss of manuscript folios, between the two of them they possess a continuous sequence of kalends and ferials extending from AD 1 to AD 644, where their joint ferial sequence ends. To illustrate the close relationship between them, the kalends, ferials and entries of the

Ƙt. ıı.

Ƙt. ııı. Paꞇꞃıcıuꞃ Ɑꞃchıepıꞃcopuꞃ eꞇ Ɑpoꞃꞇoluꞃ hıbeꞃnenꞃıum, anno aeꞇaꞇıꞃ ꞃuae cenꞇeꞃꞃımo .xxıı⁰., xuı. Ƙt. Ɑpꞃılıꞃ quıeuıꞇ, uꞇ ꝺıcıꞇuꞃ:—

O ꝫenaıꞃ Cꞃıoꞃꞇ, aıꞃem aıꞇ
Ceꞇꞃe céꝺ ꞃoꞃ caom noéaıꞇ
Ꞇeoꞃa blıaꝺna beaéꞇ ıaꞃꞃın
ꝫo báꞃ PaꞇꞃaıꞫ, Pꞃíoṁ Ɑꞃpaıꞇ.

Ƙt. ıııı.

Ƙt. u. Caꞇ Ꞇaıllꞇen ꞃoꞃ LaıꞫnıu ꞃıa Coıꞃꞃꞃe mac Néll

Ƙt. ı. Caꞇ ꞇanaıꞃı ꝫꞃaıne ın quo cecıꝺıꞇ Ꞇꞃaeé mac Pıonnchaꝺa Ꞃí LaıꞫen ꝺeꞃꞫaꞇaıꞃ. Eochaıꝺ mac Coıꞃꞃꞃe uıcꞇoꞃ ꞃuıꞇ.

Kal. ii

Kal. iii Patrick, Archbishop and Apostle of the Irish, in the 122nd year of his age, on the 16th of the Kalends of April, quievit, ut dicitur:–

Since Christ was born, a joyful reckoning,
Four hundred and fair ninety;
Three exact years after that
To the death of Patrick, Chief Apostle.

Kal. iv.

Kal. v. The battle of Taillten *was gained* over the Lagenians by Coirpre, son of Niall.

Kal. i. The second battle of Graine, in which fell Fraech, son of Finnchadh, King of Southern Leinster. Eochaidh, son of Coirpre, was the victor.

Fig. 3 Parts of pages 30–34 of William Hennessy's edition of *Chronicum Scotorum* showing the kalends 'Kł', followed by the ferial number. Under the year 'Kł. iii.' St Patrick's obit is given almost verbatim to that of the *Annals of Tigernach* (see Fig. 4). Here, the *left panel* is Hennessy's transcription from the manuscript, and the *right panel* is a transcript of his translation

Annals of Tigernach, pertaining to the period around the death of St Patrick are shown in Fig. 4. Note the similarity to Fig. 3.

For a modern reader one serious difficulty of the system of kalends and ferials is that no indication is given of the corresponding Anno Domini year with which we are familiar. Furthermore, the Roman numerals used to represent ferials are susceptible to errors when copied by scribes. For example, the three minims of 'K.iii.' were sometimes miscopied as either 'K.ii.' or 'K.iiii.'. A more serious error, pointed out by Hennessy in the citation above, arises from the representation of the numeral five in the early medieval period by 'u'; its vertical strokes are spaced identically to 'ii' and thus, when a manuscript became worn or discoloured, 'K.ii.' could be easily miscopied as 'K.u.', or vice-versa. Since from year to year the ferial number should increase by just one or two, depending on whether the first year is common or leap, errors such as these led to serious confusion. In particular, it became impossible to identify leap years.

Confronted with these transmission errors, modern scholars were unable to determine either the correct ferials, or the number of kalends, or to identify the appropriate AD year, and so they repeatedly dismissed the system. For example, in 1913 Eoin Mac Neill, having described the kalends and ferials of the third fragment of the *Annals of Tigernach*, concluded,

It follows that, of themselves, the Annals of Tigernach, in the period covered by Fragment III, afford no precise basis for dating year by year.

In 1940 Thomas O'Rahilly wrote,

as might be expected from the corrupt state of its [Tigernach's] text, they [the ferials] are often hopelessly confused. In the kindred Chronicum Scottorum, likewise the ferials are noted, irregularly and incorrectly, down to 643.

AD	1 Jan. Ferial	AT – Stokes' edition	AT – CELT Translation
490	ii	**Kl.ii.** Zeno Augustus uita decessit anno septimo mense sexto. Hi menses et sex menses Marciani addunt annum quem non numerant cronice.	**Kl.ii.** Zeno August lost his life in the seventh year and the sixth month. These months and six months of Marchan add a year which the chronicle[s] don't add.
491	iii	**K.ui.** Anastaisius regnauit annos xxuiii. PATRICIUS ARCIEPISCOPUS ET APOSTOLUS Hibernensium anno etatis sue centisimo uigessimo .xui. die Kl. Aprilis quieuit. O genemain Crist ceim ait, cethri cét íor caemnochaid, teora bliadna sáera iar soin, co bass Patraic primapstail.	**K.ui.** Anastasius ruled for 28 years. PATRICK ARCHBISHOP AND APOSTLE of the Irish in the hundred and twentieth year of his life, on the 16th [Kalends of] April rested. From Christ's birth, a pleasant step, four hundreds on fair ninety, three noble years after that to the death of Patrick the chief apostle.
492	iiii	**K.uii.** Trasamundus Vandalus Vandalorum rex catolicas eclesias clausit et .cc.xx. episcopos exilio misit Sardiniam.	**K.uii** Trasamundus Vandalus Vandalorum the king closed the Catholic churches and sent off 120 bishops in exile to Sardinia.
493	ui	**K.u.** Cath Sratha. Felix papa quieuit, cui successit Gelasius papa annos .iii.	**K.ui.** The battle of Srath. Felix the pope rested, to whom Gelasius papa succeeded for 3 years
494	uii	**K.uii.** Cath Taillten for Laignib ria Cairpre mac Neill.	**K.uii.** The battle of Tailtiu gained over Leinster by Cairbre, son of Niall.
495	i	**K.i.** Cath tanaiste Graine, in quo cecidit Fraech mac Fidchadha rí Laigen Desgabuir, la hEochaigh mac Cairpri uictor fuit.	**K.i.** The second battle of Grane, wherein Fraech, son of Fidchadh, king of south Leinster, fell by Eochaid, son of Cairbre. Eochaid was victor.

Fig. 4 *Third column*: the *Annals of Tigernach* transcribed from pp. 81–82 of Whitley Stokes' edition of the manuscript Bodleian, Rawlinson B. 488, folio 7r. The kalends 'K' or 'Kl' are followed by ferial numbers. Under the year 'K.ui.' St Patrick's obit is given, followed by a quatrain in Irish. *Fourth column*: Stokes' text translated to English following from the CELT website at http://www.ucc.ie/celt/published/T100002A/index.html. First and *second columns*: the six kalends synchronized to the AD years 490–495, followed by the true 1 January ferials for these years

In 1942, Paul Walsh wrote,

The ferial numbers are partially supplied in the early Christian portion of the so-called Annals of Tigernach and in the Chronicum Scotorum. In the latter especially, they are almost entirely wrong. They are not much better in the other compilation.

Correcting the Kalends and Ferials

In 1995, having observed the correspondence between the use of kalends and ferials to identify each year of the Padua Easter table, and those in *Annals of Tigernach* and the *Chronicum Scotorum*, I found myself confronted by the following question. Why, if these ferials were 'hopelessly confused', was Dubhaltach Mac

Fhirbhisigh, the scribe of *Chronicum Scotorum*, still copying them in circa 1640? I concluded rather that the ferial numbers must possess some useful characteristic which modern scholars had overlooked. Accordingly, I undertook an examination of their properties.

Already, from a short while after its introduction by Julius Caesar in 46 BC, the Julian year consistently followed a sequence of three common years, each comprising 365 days, followed by a leap, or bissextile, year of 366 days. Since a 365-day year is composed of 52 seven-day weeks plus a single day $[365 = (52 \times 7) + 1]$, it follows that after a common year the ferial number of 1 January must increase by one. However, if the resulting ferial number should exceed K.uii., it has to be reset to the first day of the week, i.e. to K.i. In the case of a bissextile year, the 366 days comprise 52 weeks plus 2 days $[366 = (52 \times 7) + 2]$. Therefore the ferial number of the following year must increase by two. Again, should the result exceed K.uii., it has to be reset accordingly. In Table 2, I tabulate the outcome of these two mechanisms, and we observe that the tabulation results in a 28 year sequence which simply repeats itself thereafter.

Arranged as an array of four rows and seven columns in this manner, a useful numerical structure emerges. Proceeding year by year down each column the ferial number increases by one. On moving to the start of the next column of 4 years, following the leap year, the ferial number increases by two. This pattern repeats for each of the seven columns over a 28 year cycle, so that the table may be indexed by taking the positive remainder of the Anno Domini on division by 28. For example, AD $491 = 17 \times 28 + 15$, and the 15th year of the table shows that the ferial on 1 January 491 is K.iii. Furthermore, if we move forward in time along any row, that is advancing by intervals of 4 years in each instance, the ferial number either increases by five, or, if the result would exceed K.uii., it decreases by two. For example, the ferials of the years of the second row are K.i., K.ui., K.iiii., K.ii., K.uii., K.u., K.iii., and this sequence recurs every 28 years. The other three rows, *including that of the bissextile years*, contain this identical sequence with just a different starting year. Therefore, if we compare the ferial of any given year with that 4 years earlier or 4 years later they should differ either by two or five. Moreover, *it is not necessary to know whether the year in question is common or bissextile*. This is the crucial property of ferial numbers that emerges from tabulating the 28 year cycle in the above manner.

Table 2 The 28 year cycle of the ferials of the kalends of January (1 January) arranged as seven columns of 4 years, with the bissextile (leap) years highlighted in bold

Years	1–4	5–8	9–12	13–16	17–20	21–24	25–28
Common	K.uii.	K.u.	K.iii.	K.i.	K.ui.	K.iiii.	K.ii.
Common	K.i.	K.ui.	K.iiii.	K.ii.	K.uii.	K.u.	K.iii.
Common	K.ii.	K.uii.	K.u.	K.iii.	K.i.	K.ui.	K.iiii.
Bissextile	**K.iii.**	**K.i.**	**K.ui.**	**K.iiii.**	**K.ii.**	**K.uii.**	**K.u.**

With this insight it was possible to compare each ferial of the *Annals of Tigernach* and the *Chronicum Scotorum* with that 4 years earlier and 4 years later without needing to consider whether these years were common or bissextile. This enabled the systematic detection and correction of all of the kalend and ferial errors, even in those cases where a scribe had mistakenly omitted a kalend, or had interpolated an additional kalend, or had introduced a substantial ferial error. This was published in 1998 in the *Proceedings of the Royal Irish Academy* as a table giving the manuscript reading of the kalend and ferial from the Clonmacnoise group manuscripts, aligned appropriately with the ferial cycle shown in Table 2 (Mc Carthy 1998a, pp. 245–250).

The scribal errors found in the ferial sequences of the Clonmacnoise group of annals are themselves of interest in that shared errors point to a common earlier source, whereas unique errors are most likely the result of a transcription error later than the common source. To obtain an idea of this we compare their performance over a substantial number of years preserved in common. The *Annals of Tigernach* and *Chronicum Scotorum* both extend without a break from AD 488 to AD 644, where their joint ferial sequence concludes. Over these 157 years the comparison process for the *Annals of Tigernach* led to the correction of 37 ferial errors, the restoration of nine missing kalends, and to the deletion of one interpolated kalend. In *Chronicum Scotorum* it resulted in the restoration of 55 omitted ferials, the correction of 18 ferial errors, the restoration of four missing kalends, and the deletion of six interpolated kalends (Mc Carthy 1998a, p. 219). While there are some shared ferial errors the majority are unique to each source, indicating that each has travelled a significantly different transmission pathway.

This contrast may be observed even in the small sample of years given in Figs. 3 and 4, where in the second and third years *Chronicum Scotorum* has preserved the correct ferials, 'Kl.iii.' and 'Kl.iiii', which the *Annals of Tigernach* have corrupted to 'Kl.ui.' and 'Kl.uii' respectively. On the other hand *Chronicum Scotorum* has lost the fifth kalend and its ferial, both of which the *Annals of Tigernach* have correctly preserved as 'K.uii.'. At the fourth year, however, both sources have transmitted the corrupt 'K.u.', rather than the correct 'K.ui.'. The nature and number of the scribal errors in these kalend and ferial sequences imply that both sources have undergone repeated transcription, and hence suggest their relative antiquity.

When these corrections have been made to the ferial numbers we find that the original ferial sequence of the 6 years of Fig. 4 was Kł.ii, Kł.iii, Kł.iiii, Kł.ui, Kł.uii, Kł.i., corresponding to years 14–19 of Table 2. While this ferial sequence recurs every 28 years, in this instance it may be reliably located in the last decade of the fifth century on account of the independently known chronology of the imperial and papal entries also preserved in them. For example, the emperor Anastasius' reign, registered in the *Annals of Tigernach*, is well attested to have commenced in AD 491. This consequently locates these 6 years at AD 490–5, as has been shown in the first column of Fig. 4.

When the joint ferial series of *Annals of Tigernach* and *Chronicum Scotorum* ends we may still continue collating their kalends and their entries, keeping a careful watch for those entries that can be independently assigned an AD year. What we find is a situation similar to that shown in Figs. 3 and 4, namely under each common kalend are common entries always arranged in the identical sequence. Within two decades we arrive at the common entry, 'Tenebre i callaind Mai in hora nona' (Darkness on 1 May at the ninth hour), which entry unmistakeably registers the solar eclipse of 1 May 664 that indeed traversed Ireland and Britain on the afternoon of that day (Mc Carthy and Breen 1997, pp. 122, 128–131). In this way I was able to construct a continuous series of kalends extending from the Incarnation (AD 1) to this solar eclipse in AD 664. However, when I counted the number of these kalends I found only 651, where 664 are required, implying a deficiency of 13 kalends. When I investigated the cause of this deficiency by examining all those events for which we have a reliable independent chronology, it emerged that:

• The seven kalends for the years AD 425–431 are missing.
• Six non-consecutive kalends are missing between AD 615 and AD 664.

Regarding the seven kalends missing between AD 425 and AD 431 it was straightforward to restore them. For the remaining six missing kalends, fortunately, because over AD 615–664 the *Annals of Tigernach* and *Chronicum Scotorum* preserve nine entries for which we have a reliable independent chronology, it was also possible to restore them. In this way I was able to restore the kalends and ferials of the Clonmacnoise group, and to synchronize these with the Anno Domini for AD 1–664, and for the first six centuries this structure accords with the kalend and ferial chronological tradition employed in the Padua Easter table. That 84-year Easter tradition was brought by St Columba to Iona in AD 562 where the annals were continued until circa AD 740. Furthermore, the *Annals of Tigernach* and *Chronicum Scotorum* are not beset by the ambiguity that arises from the duplicate and triplicate entries found in the *Annals of Ulster* over the early Irish Christian period.

To illustrate these points I consider the example of the year of St Patrick's death shown in Fig. 4. This places the obit for St Patrick in the year AD 491, and it is worth emphasizing that this is the only obit for him found in the Clonmacnoise group. We also note that the quatrain in Irish following the obit, most likely interpolated in the mid-eighth century, asserts that Patrick died 493 years after the birth of Christ. This thereby confirms that the poet and the annalistic interpolator both considered that Patrick had died in the last decade of the fifth century. Finally we point out that this solitary obit in *Annals of Tigernach* and *Chronicum Scotorum*, together with the correct ferial series 'Kł.ii., Kł.iii., Kł.iiii.' for AD 490–2 preserved in *Chronicum Scotorum*, contrast profoundly with the triplicate obits for St Patrick found in the *Annals of Ulster* distributed over the years AD 461–492.

Furthermore, comparing the kalends, AD years, and entries of the *Annals of Ulster* to the synchronization of the Clonmacnoise group showed that the *Annals of Ulster* are a later version, derived from a member of the Clonmacnoise group. But the chronology of the *Annals of Ulster* up to AD 663 is repeatedly distorted relative to the chronology of the Clonmacnoise group. In particular, five additional kalends were interpolated into the *Annals of Ulster* between AD 573 and AD 655, evidently in an endeavour to restore the six kalends missing between AD 615 and AD 664. It is this deficiency of one kalend that is responsible for the *Annals of Ulster's* Anno

Domini to be systematically one year too low from AD 663 to AD 1013 (Mc Carthy 1998a, pp. 251–254). Therefore, there are compelling grounds on which to consider that the chronological record of the *Annals of Tigernach* and *Chronicum Scotorum* taken together represents the best available annalistic chronological record of the early centuries of Christianity in Ireland.

In further support of this assertion I may cite the example of the date of the death of St Columba whose feast day is on 9 June, and who was known from various sources to have died in the last decade of the sixth century. However, because Adomnán (†704), in his *Vita Columbae,* written in the late seventh century, emphatically stated that Columba had died on a Sunday, his death has virtually always been located on 9 June 597, which falls on a Sunday. On the other hand, the synchronization of *Annals of Tigernach* and *Chronicum Scotorum* locates Columba's death 4 years earlier in AD 593, when 9 June fell on a Tuesday. But examination of Adomnán's account of the death of Columba betrays an awareness by Adomnán that, if the death were to fall on Sunday, then it had to be delayed by exactly 4 years. Consequently, Adomnán's *Vita* effectively confirms the accuracy of the chronology of Columba's death given by *Annals of Tigernach* and *Chronicum Scotorum.* (Mc Carthy 2014, pp. 5, 20–26).

Extension of the Synchronization

My initial collation of the kalends, ferials and entries against the years AD 1 to AD 664 was subsequently extended over the entire Christian era preserved in the *Annals of Tigernach* and *Chronicum Scotorum*, that is from AD 1 to AD 1179. This collation also included the kalends and the entries of the *Annals of Ulster* and this revealed that from AD 663 to the tenth century its kalends and entries were systematically congruent with those of the *Annals of Tigernach* and *Chronicum Scotorum*. In this way it was possible to traverse the years AD 767–803, where both the *Annals of Tigernach* and *Chronicum Scotorum* have a common lacuna. At AD 1014 these three are joined by the *Annals of Loch Cé*, which with its Connacht group companion from AD 1224, the *Annals of Connacht*, brings us up to the last kalend in the Irish annals, namely that in the *Annals of Loch Cé* for AD 1590. Since September 2000 this collation for AD 1–1590 has been available in the public domain on the Internet at www.irish-annals.cs.tcd.ie.

This extension of the collation identified further problems with the *Annals of Ulster*'s Anno Domini chronology from the late twelfth to the late fourteenth century, problems which had not been addressed in Bartholomew Mac Carthy's edition of these annals published in 1893 and 1895. It emerged that it is the kalends, ferials, and AD of the *Annals of Loch Cé* and the *Annals of Connacht* that provide a reliable chronology over the years AD 1180 to AD 1590. The *Annals of Ulster*, on the other hand, have interpolated kalends at AD 1192 and 1314, and omitted kalends at AD 1223, 1263, 1266, 1272, 1286, 1371 and 1373, and these interpolations and omissions have been registered in its Anno Domini sequence. Consequently, the

Annals of Ulster's Anno Domini chronology is only trustworthy over the years AD 1014–1191, and AD 1379–1541 (Mc Carthy 2008, pp. 349–354).

Regarding the *Annals of Inisfallen* (which are preserved in the earliest annalistic manuscript, Bodleian Rawlinson B. 503, dating from circa AD 1092), collation up to the mid-eighth century shows repeated kalend omissions and interpolations so that the editorial Anno Domini supplied by their editor, Seán Mac Airt, cannot be considered trustworthy. Finally, while the *Annals of the Four Masters* are popularly regarded as the pre-eminent annalistic compilation, sampling has shown its chronology to be highly erratic, and for this reason it has not been included in the collation (Mc Carthy 2008, p. 299).

To give the reader an idea of the organisation of this collation I reproduce in Table 3 my tabulation for the year AD 491. The headings in the first row identify the contents of each column. In the second row the first column gives the synchronized AD year and the second column gives the correct 1 January ferial for this year, 'K.iii.', then columns three, four and six show the kalends and other chronological criteria recorded in the *Annals of Tigernach*, *Chronicum Scotorum* and the *Annals of Ulster* respectively. As we have seen, the *Annals of Tigernach* give a corrupt 'K.ui' instead of the correct 'K.iii.' given in *Chronicum Scotorum*, while the *Annals of Ulster* have the correct Anno Domini for this year '.cccc.xc.i'. The third row shows that the *Annals of Tigernach*'s first entry registers the commencement of Anastasius' 28-year reign, that the *Annals of Ulster* register this reign as the second entry under its previous kalend with AD 490, and that in his *Chronica maiora* Bede registers this reign at paragraph 505 of Charles Jones' edition of Bede's chronicle. Finally, the fourth row shows that the *Annals of Tigernach*'s second entry registers Patrick's obit as 'Patricius q[ueiuit], while *Chronicum Scotorum*'s first entry is verbatim, that *Mageoghagan's Book* registers the obit in English as 'Patrick ... d[yed]', that the *Annals of Ulster* have two obits for Patrick, one as the first entry under its AD 491, the other as the fourth entry under its AD 492, that AI registers the obit as the first entry under the editorial AD 496, and that the *Annals of Boyle* register it at paragraph 162 of Martin Freeman's edition. The textual and chronological details of all of these obits show that they are cognate to some degree, while the *Annals of Ulster*'s entry at 461.2 reading 'Hic alii quietem Patrici dicunt' shows that it is a later interpolation not cognate with the other obits, and hence it is not tabulated at AD 491.

Table 3 The synchronisation of the kalends and ferials of the *Annals of Tigernach* and *Chronicum Scotorum* with the Anno Domini and the true Julian ferial for AD 491, and the collation of the events of the other annals and chronicles with these

AD	Feria	AT	CS	MB	AU	AI/AB/AR/FA/CM/AC
491	K.iii.	K.ui	Kl.iii.		Kl. AD.cccc.xc.i	
		Anastasius reg. xxuiii			490.2	C:505
		Patricius q.	Patricius q.	Patrick d.	491.1,492.4	I:496.1,B:162

Conclusions

The Irish annals constitute a very substantial and valuable collection of chronicles which, when taken together, span the period from "Creation" to the earlier seventeenth century. Their source for the earliest millennia is the Bible, followed by the Christian chronicles of the Mediterranean world up until the late fourth century. Then, with the arrival of Christianity in Ireland in the earlier fifth century, their focus moves to, and remains in, Ireland, with thousands of cryptic entries providing records of events from an Irish Christian perspective. These are predominantly accounts of Irish ecclesiastical and secular obits and conflicts, but they include occasional accounts of events in Britain and the Continent, as well as phenomenological events such as eclipses, comets, weather extremes, plagues and famines.

The evaluation of this corpus by historians of early Christian Ireland over the last century has concentrated upon that part of the *Annals of Ulster* commencing from AD 431. They have preferred the *Annals of Ulster* because its earlier entries are predominantly in Latin, or, if in Irish, these are written in an earlier orthography than that of the other annals. It is also a matter of familiarity and convenience that the *Annals of Ulster* provide an Anno Domini datum for each year. However, the perceived antiquity of the language and orthography of chronicle entries cannot provide any guarantee of the antiquity or accuracy of its chronology, that is the distribution of these entries across the chronicle years. Rather, the very appearance in the fifth century of the *Annals of Ulster* of Anno Domini years, which originate with the Dionysiac Paschal tradition that was not adopted in Iona until the eighth century, signals their retrospection. This implies that the chronological organisation of the *Annals of Ulster* is the result of a later undertaking to project Anno Domini years back to AD 431, the year of the arrival of bishop Palladius in Ireland. Furthermore, it is evident that this undertaking also involved a substantial revision of the chronology of the entries over the years AD 431–663. Moreover, the detail that from AD 664 the *Annals of Ulster*'s Anno Domini systematically trails the established chronology of events by one year up until AD 1012, where the deficiency is rectified by the simple expedient of omitting AD 1013, is a further indication of the poor competence with which these Anno Domini were inserted. The fact that from the earlier fifth to the middle seventh century the *Annals of Ulster* also contain repeated duplications of entries, and sometimes triplications, is a further warning that the chronological integrity of its events in this period has been ineptly handled.

On the other hand it is known that the earlier Paschal tradition followed in Ireland employed an 84-year Easter table, and that this 84-year Easter tradition was subsequently eclipsed by the Dionysiac tradition. This happened in Britain at the synod of Whitby in AD 664, and in Ireland in the subsequent centuries. So completely was the 84-year Easter table eclipsed that until the late twentieth century the details of its construction and organisation were quite unknown. However, the full copy of the Padua Easter table discovered in 1985 by Dáibhí Ó Cróinín revealed that the 84-year Easter tradition followed in early Christian Ireland employed as its primary sequencing mechanism a 'Kl' followed by the feria of 1 January. Since the same sequencing mechanism is used in the Clonmacnoise group of annals, where in the *Annals of Tigernach* and *Chronicum Scotorum* it jointly extends from the Incarnation to AD 644, this strongly suggested that these annals preserve an early chronological tradition.

But, because these ferials written as a Roman numeral between one and seven contain significant numbers of scribal errors, they had been dismissed by scholars as 'hopelessly confused'. However, examination of the 28 year ferial cycle revealed the invariant property that, when compared at 4-year intervals, the ferial must either increment by five or decrement by two. This knowledge enabled the systematic identification and correction of all scribal kalend and ferial errors over the entire sequence. But when the chronology of this restored sequence of kalends and ferials was compared with those events for which we have a reliable independent chronology it emerged that, following its original compilation, the chronicle had been subject to subsequent revision. Namely, the seven kalends for AD 425–431 and six non-consecutive kalends between AD 615 and AD 664 had been deleted. When these 13 kalends are restored the result appears to be our most reliable chronology for early Christian Ireland. For examples, the Clonmacnoise group obit for St Patrick is located at AD 491, preceded in the *Annals of Tigernach* by a record of the start of the reign of the emperor Anastasius, which is independently known to have commenced in that year. The Clonmacnoise group locates the death of St Columba at AD 593, which is effectively confirmed by the account of Columba's death given by Adomnán in his *Vita Columbae* compiled in circa 697.

The kalends and the entries of this restored and corrected Clonmacnoise group have been synchronized to the Anno Domini year, and this then collated with cognate entries from other annals and chronicles. This collation was carried forward to the conclusion of the Clonmacnoise group in AD 1179. From there it was continued up to the last kalend in the Irish annals at AD 1590, and the resulting collation made available at www.irish-annals.cs.tcd.ie. This continuation disclosed further Anno Domini problems in the *Annals of Ulster* from the late twelfth to the late fourteenth century, while, on the other hand, the Anno Domini of the *Annals of Loch Cé* and the *Annals of Connacht* were shown to provide a reliable basis for the chronology of the entire later medieval period.

This collation and synchronization was compiled by me over the years 1999–2005 with the primary purpose of identifying reliable annalistic chronological criteria, and constructing a dependable Anno Domini chronology for the entire Christian period. Because the chronological problems are the most complex in the earlier Christian centuries I comprehensively collated annalistic entries from the Incarnation up until AD 766, by which time one is dealing with about 10 entries per annum. In the subsequent centuries this number rises to sometimes over 50 entries per annum, and this was far beyond my capacity to document. However, it would be a most desirable objective to comprehensively collate and to synchronize references to all of the annalistic entries. For this would both provide a unified chronology for all of the Irish medieval annalistic sources, and facilitate the textual and semantic comparison of all annalistic entries referring to the same event. Furthermore, full collation will highlight those entries unique to each annalistic source, and these entries provide a valuable index to each source's particular history and origin. Since electronic editions of the principal annalistic sources have been made available on the invaluable CELT (Corpus of Electronic Texts) website at the National University of Ireland, Cork, this substantial collation task would be susceptible to a fair degree of automation, and this would make an excellent collaborative project involving information technology and Irish medieval history. I may add that, because of their chaotic chronology I omitted the *Annals of the Four Masters* from my collation, but it would also be desirable that their vast collection of entries should be included in any such expansion of the collation.

Acknowledgement I wish to gratefully acknowledge Ralph Kenna's enthusiasm and generosity in enabling me to contribute to the Maths Meets Myths workshop held in the University of Coventry on the 11–12 September, 2014. I am also indebted to him for his numerous very constructive suggestions made in the course of the drafting of this article.

Bibliography

Comyn, D., & Dinneen, P. (Eds.), *Foras Feasa ar Éirinn le Seathrún Céitinn: The history of Ireland* i–iv, Irish Texts Society 4, 8, 9, 15 (London 1902, 1908, 1908, 1915).

Dumville, D. N. (1982). Latin and Irish in the annals of Ulster, A.D. 431–1050. In D. Whitelock, R. McKitterick, & D. N. Dumville (Eds.), *Ireland in early medieval Europe: Studies in memory of Kathleen Hughes* (pp. 320–341). Cambridge: Cambridge University Press.

Etchingham, C. (2002). Les Vikings dans les sources documentaires irlandaises: le cas des annals. In É. Ridel (Ed.), *L'Héritage maritime des Vikings en Europe de l'Ouest* (pp. 35–56). Presses Universitaires de Caen, (Caen, 2002).

Freeman, A. M. (Ed.) (1944). *Annála Connacht: The Annals of Connacht (A.D. 1224–1544).* (Dublin 1944, repr. 1970).

Freeman, A. M. (Ed.), The annals in Cotton MS Titus A xxv. *Revue Celtique* 41 (1924) 301–30; 42 (1925) 283–305; 43 (1926) 358–84; 44 (1927) 336–61. Otherwise known as the *Annals of Boyle.*

Gleeson, D., & Mac Airt, S. (Eds. & Trans.) (1959). The annals of Roscrea. *Proceedings of the Royal Irish Academy* 59C, 137–180

Hennessy, W. M. (Ed. & Trans.) (1866). *Chronicum scotorum. A chronicle of Irish Affairs, from the earliest times to A.D. 1135, with a supplement containing the events from 1141 to 1150.* Roll Series 46. London, 1866. Reprinted: Wiesbaden, 1964. Available online from https://ia700409. us.archive.org/5/items/chronicumscotoru00macfuoft/chronicumscotoru00macfuoft.pdf

Hennessy, W. M. (Ed. & Trans.) (1871) *Annals of Loch Cé: A chronicle of Irish affairs, 1014–1690* i–ii (London 1871, repr. Dublin 1939).

Hennessy, W. M. (Ed. & Trans.) (1887). *Annála Uladh: The Annals of Ulster* i (Dublin 1887, 1999). Reproducing AD 431–1056.

Jaski, B., & Mc Carthy, D. (2012). *The annals of Roscrea: A diplomatic edition.* Roscrea: Roscrea Heritage Society.

Jones, C. W. (Ed.) (1977). 'Chronica maiora' in *Bedae venerabilis opera*, CCSL cxxiii B (Turnhout 1977) 461–535.

Mac Carthy, B. (Ed. & Trans.), *Annála Uladh: The Annals of Ulster* ii–iv (Dublin 1893, 1895, 1901, 1999). Vol. ii–iii reproduce AD 1057–1588 with some lacunae, and vol. iv contains the Introduction and Index.

Mac Airt, S., & Mac Niocaill, G. (Eds. & Trans.) (1983) *The Annals of Ulster (to A.D. 1131)*. Dublin: Dublin Institute for Advanced Studies. Reproduces AD *c*.81–1131.

Mac Airt, S (Ed.), 'Middle-Irish poems on World-kingship', *Études Celtique* vi (1953–4) 255–81, vii (1955–6) 18–45. The poems of Flann Mainistreach.

Mac Airt, S (Ed. & Trans.) (1951) *The Annals of Inisfallen (MS Rawlinson B 503)*. Dublin: Dublin Institute for Advanced Studies.

Macalister, R. A. S. (Ed.) (1938–56). *Lebor Gabála Érenn* i–v in ITS vols. xxxiv, xxxv, xxxix, xli, xliv. Dublin: Irish Texts Society.

Mac Niocaill, G. (1975). *The Medieval Irish annals* (Medieval Irish History Series, Vol. 3). Dublin: Dublin Historical Association.

Mc Carthy, D. P. (1998a). The chronology of the Irish Annals. *Proceedings of the Royal Irish Academy, 98C*, 203–255.

Mc Carthy, D. P. (1998b). The status of the pre-Patrician Irish annals. *Peritia, 12*, 98–152.

Mc Carthy, D. P. (1999–2005). *Chronological synchronisation of the Irish Annals*, at www.irish-annals.cs.tcd.ie

Mc Carthy, D. P. (1999–2005). *Collation of the Irish regnal canon*, at www.irish-annals.cs.tcd.ie

Mc Carthy, D. P. (2000). The chronology of St Brigit of Kildare. *Peritia, 14*, 255–81.

Mc Carthy, D. P. (2001). The chronology and sources of the early Irish annals. *Early Medieval Europe, 10*, 323–41.

Mc Carthy, D. P. (2008). *The Irish Annals: Their genesis, evolution and history*. Dublin: Four Courts Press, repr. 2010.

Mc Carthy, D. P. (2014). The chronology of Saint Columba's life. In P. Moran & I. Warntjes (Eds.), *Early medieval Ireland and Europe: Chronology, contacts, scholarship. A Festschrift for Dáibhí Ó Cróinín* (pp. 3–32). Turnhout: Brepols.

Mc Carthy, D. P., & Breen, A. (1997). An evaluation of astronomical observations in the Irish annals. *Vistas in Astronomy, 41*(1), 117–138.

Mc Carthy, D., & Ó Cróinín, D. (1990). The lost Irish 84-year Easter table rediscovered. *Peritia* 6–7 (1987–8) [1990] 227–242.

Murphy, D. (Ed.) (1896). *The annals of Clonmacnoise, being annals of Ireland from the earliest period to A.D.1408, translated into English, A.D.1627 by Conell Mageoghagan*. Dublin, 1896, repr. 1993.

O'Donovan, J. (Ed.) (1841–1851). *Annala Rioghachta Eireann: Annals of the Kingdom of Ireland* i–vii (Dublin; repr. New York 1966).

Smith, P. J. (Ed.). (2007). *Three poems ascribed to Gilla Cóemáin: A critical edition of the work of an eleventh-century Irish scholar*. Münster: Nodus-Publ.

Stokes, W. (Ed. & Trans.). *The Annals of Tigernach*, first published in *Revue Celtique* 16 (1895) 374–419; 17 (1896) 6–33, 119–263, 337–420; 18 (1897) 9–59, 150–97, 267–303. The facsimile edition *The Annals of Tigernach* i–ii (Felinfach, Wales 1993) is paginated as 1–223, 224–466.

Todd, J. H. (Ed.). (1867). *Cogadh Gaedhel re Gallaibh*. London: Longmans.

Mapping Literate Networks in Early Medieval Ireland

Quantitative Realities, Social Mythologies?

Elva Johnston

Abstract The medieval Irish chronicles are documentary sources of the highest importance. They provide contemporary records of major political events from the late sixth century AD. Each entry is anchored in a wealth of place-name references and, literally, thousands of death notices, providing a dataset whose scale is suitable for the application of network theory. This chapter will take one prominent group from the chronicles as a case study, the literate elite who produced most of our sources. It will explore how network theory can enhance our understanding of them, and will also consider the extent to which their networks are mirrored in the literary sources in which they feature. How are their social networks depicted in narrative texts? Do these appear to be realistic or are they idealized? By considering one definable group, the literate elite, this chapter aims to provide a framework through which network theory can be more widely and usefully employed.

Identifying the Dataset

The early medieval Irish chronicles, usually referred to as annals, are key documentary sources for Irish history (Grabowski and Dumville 1984; Mc Carthy 2008; Evans 2010). They provide contemporary year-by-year accounts of events from the late sixth century AD through to, and beyond, the arrival of the Anglo-Normans in the twelfth century. The typical entry is short, even laconic, noting a battle, an outbreak of disease or death. There is little in the way of explicit editorialising comment, making many entries seem spare. Yet, the chronicles are no mere dateline: they are replete with place-name references, geographically anchoring thousands of death notices. These death notices, called obits in the secondary literature, are a primary way in which information is encoded. The vast majority of them commemorate members of the Irish elites, ecclesiastical, professional and secular. These obits have been cross-referenced, although not scientifically and only on a

E. Johnston (✉)
School of History, University College Dublin, Dublin, Ireland
e-mail: elva.johnston@ucd.ie

© Springer International Publishing Switzerland 2017
R. Kenna et al. (eds.), *Maths Meets Myths: Quantitative Approaches to Ancient Narratives*, Understanding Complex Systems,
DOI 10.1007/978-3-319-39445-9_11

case-by-case basis, with the very substantial medieval Irish genealogical corpus. The published genealogies contain the names of more than 12,000 individuals who lived in Ireland before the twelfth century, a figure that is more than doubled when the unpublished tracts are also considered (O'Brien 1962; Ó Corráin 1998: 180–181). The most common genealogical structure is an agnatic patriline with women featuring rarely (Johnston 2013: 79–89), even though their irreplaceable role within kinship networks was well understood (Kelly 1988: 14–15; Charles-Edwards 1993: 76–77; Bitel 1996: 140–148). The chroniclers' bias mirrors the basic hierarchical and patriarchal nature of early medieval Irish society (Kelly 1988: 68–79). The women who do appear are frequently either saints or their ecclesiastical successors. Others are royal women, especially those accounted queens of Tara (Connon 2000: 99–101).

Thus, the chronicles, supplemented by the genealogies, provide a dataset whose scale makes it amenable to the application of quantitative methods and, indeed, network science. Each separate individual is a node whose direct and indirect links to other nodes can be mapped (Mac Carron and Kenna 2012; MacCarron and Kenna 2014). Some nodes seem isolated while others are embedded in a complex series of links and are heavily connected, depending on the degree to which their activities are recorded in the chronicles. Significantly, the chroniclers emphasize the connectivity of Irish society, expressed through the actions of particular individuals. These actions are constrained by factors such as dynastic membership, geographical powerbase and an ability to create alliances. In a society where kingship had a strong conceptual underpinning but a relatively weak bureaucratic footprint (Charles-Edwards 2000: 522–585; Doherty 2005), the alliances created through manipulation of ecclesiastical and political networks were paramount in maintaining and building power. They created a thick clustering around influential men, such as kings and heads of churches. It is the shifting nature of these aristocratic networks which sometimes makes early medieval Irish political history appear forbidding to the non-specialist. Its essential dynamic is expressed through a highly developed and nuanced connectivity; its public world is purposefully embedded in the personal, but is no less organized as a result.

The medieval Irish chronicles, then, are rich sources. However, they are not amenable to presentation in a single paper due to the attendant complexities. These include the scale of the dataset as well as the different areas of concern emphasized by the monastic chroniclers, all of which need to be unpicked. For instance, the chroniclers show an obvious interest in secular and ecclesiastical politics; they also note natural phenomena such as outbreaks of disease in humans and animals, astronomical data and unusual weather events (Mc Carthy and Breen 1997; Mc Cormick et al. 2007). The result is a layering of data which can be cross-referenced and collated on a minute scale, something which would require a major project. This paper, therefore, will take one particular area and will argue that the application of network theory can greatly enhance our understanding of it, serving as a case study for a wider field of potential investigation. The focus will be on the literate elite, especially ecclesiastics, who produced the surviving sources and whose interests played a major role in the compilation of the early medieval chronicles.

The chroniclers use a stock vocabulary to identify these literate professionals, including Latin terms such as *scriba* 'scribe', *sapiens* 'exegete' and *doctor* 'doctrinal expert' as well as the vernacular *fer léigind* 'master of monastic school' and *suí* 'scholar' (Johnston 2013: 92–130). The meaning of words fluctuated: for instance a *scriba* might be a scribe or a canon lawyer (Ó Corráin et al. 1984: 398–399; Johnston 2013: 120–124), but there is broad consistency in chronicle usage which allows for comparisons between texts and across time. Moreover, the chronicles frequently identify the ecclesiastical and familial associations of their subjects, especially after 900 when entries become longer and more informative (Dumville 1982: 328–341). These elements, personal name, place of residence and family membership, are the basis of a social network. Not only can individuals be considered as nodes but so can their institutional bases, establishing a socio-geographical framework for situating their interactions. No such comprehensive study has ever been undertaken, although preliminary elements of it underpinned my monograph analyzing the role of literacy and its practitioners within early medieval Irish society (Johnston 2013). During its research, I compiled a database of the literate elite, based on annal entries extending from AD 797–1002, which will be the basis for much of the present study, although other relevant material will be considered when necessary. This consisted of 198 separate named individuals (Johnston 2013: 177–202). The paper will also consider the extent to which this network is mirrored in literary sources which feature actual members of the literate elite. In these narrative texts how are their social networks depicted? Are they realistic or idealized (Mac Carron and Kenna 2012: 5–6)? Do they reflect the way these people imagined their own role in society?

The Early Medieval Irish Chronicles: Origins, Biases and Value

The starting point is an understanding of the early medieval Irish chronicles, their origins, functions and in-built biases. This will establish the evidential base from which further questions can be asked. There are several extant medieval Irish chronicles, none of which survive in a contemporary manuscript. Moreover, material has been lost in the process of redaction. Nevertheless, a degree of scholarly consensus has emerged concerning the origins and evolution of Irish chronicle writing. This consensus holds that a no-longer extant text, dubbed the Chronicle of Ireland, underlies all the early medieval Irish chronicles until the tenth century (Hughes 1972: 99–159; Evans 2010: 115–170). It is believed to have originated in Iona in the latter half of the sixth century before eventually being continued in an ecclesiastical centre in the midland kingdom of Brega from *c.* 740–911 (Charles-Edwards 2006 vol. 1: 9–16). The chronicle was written in a combination of Irish and Latin, with Irish beginning to predominate from the ninth century (Dumville 1982). It has been long believed that the chronicle known as the Annals of Ulster is its best surviving witness. Fortunately, this text has been transmitted with such a high degree of accuracy that it provides valuable evidence for diachronic linguistic developments in Irish, despite the relatively late date of its most important surviving manuscript (Ó Catháin 1933; Dumville 1982). This accuracy is critical as, despite elements of retrospective revision and interpolation from other sources, the Chronicle of Ireland core is relatively easily distinguished. Thus, the Annals of Ulster afford a glimpse of unfolding events from the late sixth century. Literate networks can be examined as dynamic entities, evolving across several centuries.

The evidence diversifies considerably after AD 911–912. At this point a Clonmacnoise Chronicle or, perhaps, a set of related chronicles, diverged from the Chronicle of Ireland (Grabowski and Dumville 1984: 45–55; Evans 2010: 67–72). This has several important witnesses, particularly the Annals of Tigernach. Unfortunately, the extant text is incomplete. For instance, it has a major lacuna extending from AD 766–974 and abbreviates much of the ecclesiastical material which must have been present in its original source (Grabowski and Dumville 1984: 163–164; Evans 2010: 57–62). This is frustrating as clerics formed a large proportion of the literate elite. Tigernach can be supplemented by other chronicles, especially *Chronicum Scotorum* with which it shares a common source (Evans 2010: 50–57) but it too has gaps, including one between AD 723–803. There are other less important texts drawing on the Clonmacnoise Chronicle, including the Annals of Roscrea, the English-language Annals of Clonmacnoise and the Fragmentary Annals of Ireland. The most significant chronicle, besides the Annals of Tigernach and *Chronicum Scotorum*, which forms part of this Clonmacnois group of texts, is the Annals of Inisfallen. From the tenth century, when its entries become more detailed, it consistently references events in Munster and appears to incorporate unique sources from that region (Grabowski and Dumville 1984: 73). In broad outline then, the early medieval chronicle evidence becomes richer over time, especially following the break in the texts after 911. This clearly has implications for the depth and complexity of possible network modelling.

This picture has been considerably complicated and challenged by the work of Daniel Mc Carthy. He has suggested a somewhat earlier horizon for the beginning of chronicle writing (Mc Carthy 2008: 159–163). More fundamentally, he has disputed the centrality of the Annals of Ulster and has suggested that historians have been too reliant upon it for establishing their chronology of Irish history (Mc Carthy 1998). He has identified the Clonmacnoise Chronicle group as the key, arguing that the Iona Chronicle, which underpinned the Chronicle of Ireland, was continued in Moville from the 740s and in Clonmacnoise from 752 (Mc Carthy 2008: 168–197). In addition, he has questioned the consensus which has identified 911 as the point at which the Chronicle of Ireland ends; on the contrary, he argues that the Annals of Ulster are dependent on the Clonmacnoise Chronicle up to the mid-tenth century (Mc Carthy 2008: 61–117). Mc Carthy's re-evaluation of the relationship between the early medieval chronicles has stimulated several recent studies devoted to reconsidering the Chronicle of Ireland hypothesis (Evans 2010: 67–72; Flechner 2013). In the main, these have reaffirmed its position as the base text for both the Annals of Ulster and the Clonmacnoise group up to 911, with the Annals of Ulster retaining its central position. On the other hand, Mc Carthy's convincing reconstruction of the chronological apparatus of the Annals of Tigernach and *Chronicum Scotorum*, one which shows it to be superior to that of the Annals of Ulster at several points, does indicate that relying on the latter alone is unwise (Mc Carthy 1998).

In any case, there are other reasons for not turning solely to that text, ones which relate to its definite geographical bias. All of the early medieval chronicles prioritize events within their own region, even though they share an implicit framework that ostensibly embraces the entire island. So, for example, the Annals of Ulster have a strong interest in Armagh and the midland kingdom of Brega, during the ninth and tenth centuries. This reflects the political influence of the powerful Patrician ecclesiastical federation, which looked to Armagh as its head, as well as pointing to the likelihood that the Chronicle of Ireland was being continued in an ecclesiastical centre in Brega during the period (Evans 2010: 17–44). This focus excludes centres that were key nodes in literate networks, skewing any potential mapping of the connections between them. Furthermore, Armagh is almost certainly over-represented. Of the 92 literate professionals recorded in the Annals of Ulster during the ninth and tenth centuries, 16 % are directly associated with Armagh. This bias in the data can be partly offset through examining the other early medieval chronicles, especially after 911 when the Clonmacnoise group diverges from the Chronicle of Ireland core. These chronicles have a different, although overlapping, geographical slant to the Annals of Ulster. As a result, while they share many events and death notices, there is a far greater emphasis on the hinterland of Clonmacnoise (Evans 2010: 60–62). This major monastery dominated the Shannon basin and its influence stretched into the midlands, north Munster and Connacht (Kehnel 1995: 90–132; ÓFloinn 1998). This extensive influence is reflected in the Clonmacnoise group of chronicles. Unfortunately, the various breaks in these texts, combined with the abbreviated nature of the Annals of Inisfallen until the tenth century, does mean that the record lacks fullness. It can be supplemented by the Fragmentary Annals which are based, at least in part, on the Clonmacnoise Chronicle (Mac Niocaill 1975: 24; Radner 1999) and may well have originated in Clonenagh, a church which lay within the Leinster border kingdom of Loígis (Radner 1999: 520–525; Downham 2004). Therefore, despite difficulties, this family of chronicles significantly extends the geographical range of early medieval Irish literate networks. For instance, a quarter of the entries devoted to the ecclesiastical literate elite feature churchmen based in Munster, a striking contrast with their paucity in the Annals of Ulster. Moreover, they show a far greater interest in the secular literate classes than the Annals of Ulster (Johnston 2013: 139–140), allowing for a deeper understanding of how these networks functioned.

Two concrete examples demonstrate these factors in operation. The first is a well-known professional *fili* or poet, Urard mac Coise (†990), who may be the author of the semi-autobiographical tale *Airec Menman Uraird Meic Coise* (Byrne 1908; Mac Cana 1980: 33–38). His prominence is such that he is one of only two poets commemorated in the Annals of the Ulster in the tenth century. Urard's death notice in this text simply states that he was the chief poet of Ireland. *Chronicum Scotorum* and the Annals of Tigernach add significant information, remarking that the *fili* died in penance at Clonmacnoise, immediately placing him in the proximity of an important ecclesiastical network. This proximity is further strengthened in *Airec Menman Uraird Meic Coise* where the poet is supported in his dealings with Domnall mac Muirchertaig (†980), Uí Néill overking of Tara, by Flann, *fer léigind* of Clonmacnoise. The *fer léigind* was a high-status literate professional associated with teaching (Johnston 2013: 124–128). It seems certain that this Flann is identical with Flann mac Maíle Michíl (†979) who was a contemporary of Urard and *fer léigind* at Clonmacnoise. The extra information in the Clonmacnoise Chronicle-derived texts allows a simple network to be constructed around Urard, one which links him directly with a high-ranked ecclesiastic and the most important king in Ireland. The latter, Domnall mac Muirchertaig, was the primary node at the centre of the extensive Uí Néill system of political alliances. In addition, according to *Chronicum Scotorum*, Flann was the head of the church of Cluain Deochra (Cloneogher) in modern County Longford, a region where Clonmacnoise exerted considerable sway. Urard personally connects two extensive geo-political networks, one secular and the other ecclesiastical. None of this could be argued with any certainty in the absence of the Clonmacnoise Chronicle evidence.

The second example is Snédgus (†888), an ecclesiastical scholar who is commemorated in *Chronicum Scotorum*, but not in the Annals of Ulster. *Chronicum Scotorum* states that Snédgus was a *sapiens*, an exegete (Ireland 1996; Johnston 2013: 102–112) based in Castledermot in northern Leinster. This church was a significant site whose foundation in 802 by Díarmait ua Áedo Róin (†825) was recorded by the Annals of Inisfallen. Furthermore, *Chronicum Scotorum* also notes that Snédgus was the *aite* 'foster-father' of Cormac mac Cuilennáin, a powerful king of Munster, as well as a bishop, who died at the Battle of Belach Mugna in 908 (Ó Corráin 1972: 111–114; Russell 2004: http://www.oxforddnb.com/view/article/6319). In this context, the term *aite* indicates that Snédgus was the mentor and teacher of the young Cormac. It is almost certain that this information was added to the entry, sometime after the death of Cormac, as the future king was likely a minor figure in 888. In fact, scholars have long speculated that he rose to prominence as a compromise candidate with his episcopal status playing a central role (Byrne 1973: 214). Cormac was a member of the ruling dynasty in Munster, but came from an obscure branch. The presence of clerical rulers in Munster during the ninth and tenth centuries has appeared *sui generis* (Byrne 1973: 211–215). However, it is likely explained by the inter-connections between the secular and ecclesiastic. Cormac was a well-placed node on both networks and this boosted his chance at kingship during a period of significant dynastic uncertainty. Moreover, Cormac's background helps explain his political trajectory, especially his interest in controlling Leinster which led, ultimately, to his death. After all, it is probable that he spent his formative years in a Leinster monastery.

Urard and Snédgus show the potential for considering the early medieval chronicles as, fundamentally, accounts of networks and their interactions. A final source remains to be considered, one which greatly expands the dataset but which also presents particular challenges. This is the Annals of the Four Masters, a seventeenth-century chronicle, composed in annalistic style, which looks back to its medieval forebears. The chronicle was the product of a major collaborative effort during the 1620s–1630s, organized by the Irish Franciscans, particularly Mícheál Ó Cléirigh. Bernadette Cunningham has shown that the chronicle is a work of total history, outlining the story of the Irish, of their saints and of their kings, from the island's first habitation down to its seventeenth-century present. The chronicle is imbued with the politics of the Counter-Reformation and one of its aims was to situate Ireland among the European nations (Cunningham 2010: 31–33, 176–214). The resulting work is one of remarkable depth; it is far greater in extent and detail than any of the surviving medieval chronicles. Unlike the latter, the Annals of the Four Masters are retrospective, created in a specific short time-frame for particular reasons. As a chronicle, it provides an interpretative overview, not a rush of contemporary record. However, this retrospective nature is what makes it so useful for a study of the early medieval past. Mícheál Ó Cléirigh relied heavily on the early medieval chronicles, which he frequently cross-referenced with genealogies and king lists. An analysis of entries in the Annals of the Four Masters reveals the extent to which they systematically incorporate the extant sources. Moreover, Ó Cléirigh had access to better versions of the Annals of Ulster and the Annals of Tigernach than now survive as well as texts which are lost (McGowan 2004: 3–4; Cunningham 2010: 46–49). When Ó Cléirigh and his sources can be directly compared, the result is one of impressive fidelity (Johnston 2013: 178–196). The Annals of the Four Masters operate on the basis of accumulation and harmonisation rather than replacement or invention. The patterns of reinterpretation are also relatively consistent. For instance, Ó Cléirigh frequently replaced the early medieval use of *rí* and *rex* for a multitude of kings, ranging from the paltry to the powerful, with words such as *tigerna* 'lord' to create the impression of an ancient and unitary Irish kingdom (McGowan 2004: 7–11, 21–24). Similarly, the text applies semantically broad terms such as *ecnaid* 'wise man' or 'learned man' where the early medieval chronicles are more specific. It seems highly unlikely that the wide geographical span of the Annals of the Four Masters is a construct. Used carefully, in conjunction with the early medieval chronicles upon which it draws, the text greatly enhances the available dataset.

Evolving Networks of Literacy and Power

Having established that the medieval Irish chronicles, supplemented by the Annals of the Four Masters, are best understood as reflecting the actions of deeply connected political and ecclesiastical networks, it remains to consider, more fully, how these intersected. As stated previously, the totality of the dataset is too large for a single paper; therefore, the focus will remain on the literate elite during the ninth and tenth centuries. Scholars have long appreciated that the relationship between church and society was a core aspect of how the early medieval Irish viewed their world. There has been a strong inclination to analyze this relationship through the medium of vernacular law (Ó Corráin et al. 1984). While very useful, this has had the unintended consequence of downplaying the significance of other sources such as the chronicles. It is believed that the earliest legal texts date from the second half of the seventh century (Breatnach 2011), with most of the corpus redacted by the end of the ninth. However, the tradition was not static: it accumulated gloss and commentary, both of which allowed for the on-going development of legal ideas, albeit within schematic hierarchical structures. On the other hand, it is worth remembering that the law tracts are not evolving documents in the same way as the chronicles. In some cases, such as the provision of sick maintenance, they preserve the memory of institutions which were already at vanishing point during the period of their redaction (Watkins 1976; Kelly 1988: 130–131). As a whole, these texts emphasize the high status of literate churchman within society, but in a general rather than individual manner. For example, the important eighth-century tract known as *Bretha Nemed Déidenach* states that the ecclesiastical scholar, judge and poet were the three officials necessary for any *tuath*, or petty kingdom (Gwynn 1940: 31). These *tuatha* were small scale and ubiquitous, with estimates suggesting that there may have been upwards of a hundred of them in existence (MacCotter 2008: 41–44). *Bretha Nemed*'s stipulation creates an idealized network, with scholar, judge and poet as nodes connected to the king of the *tuath*. This is similar to Urard mac Coise's network, but does not have the same potential for extrapolating the real links between different nodes. Nevertheless, the legal evidence underlines the fact that a wide geographical distribution of literacy was considered normative. On the surface, this finds support in the medieval chronicles: they record nearly 70 different ecclesiastical centres associated with literacy in the ninth and tenth centuries (Johnston 2013: 196–198). However, during the course of the latter part of the ninth century and especially the tenth, this changed, with large centres coming to dominate. This is more easily traced through the chronicles than legal sources. The diachronic nature of the chronicles takes them beyond schematisation.

The key factors appear to be socio-political. From at least the end of the eighth century the Irish overkings had gradually accumulated more power (Ó Corráin 1978: 23–35; MacCotter 2008: 23), with the result that *tuatha* became less politically significant. There are a multitude of reasons for this, including economic developments and responses to Viking raiding, trading and settlement. These trends coincide with a noticeable concentration and consolidation of influence among a few major churches such as Armagh and Clonmacnoise (Johnston 2013: 67, 92–130). Furthermore, the rise to prominence of monastic schools, associated with *fir léigind*, ensured that high-ranking ecclesiastical literate professionals came to be increasingly linked with these institutions. Of course, such ecclesiastics had always been politically significant. The career of Adomnán (†704), abbot of Iona, is indicative (Ní Dhonnchadha 2004: http://www.oxforddnb.com/view/article/110). Adomnán was so well-connected that he was able to call on an impressive number of royal and ecclesiastical supporters as guarantors for his *lex innocentium*, a legal ordinance aimed at protecting non-combatants during warfare (Ní Dhonnchadha 1982). How did later literate churchmen respond to new circumstances? Once again, network theory provides insight. The change in distribution of literate professionals can be visualized as a network. Its top layer is composed of a small number of major institutional nodes, such as Armagh and Clonmacnoise. These were connected to a second layer, made up of very many dependent churches. Basically, a horizontal network, composed of independent and semi-independent centres, had evolved into one which was both vertical and horizontal. Some foundations, such as Armagh, had developed a vertical pyramidal network at an early point but the trend certainly accelerated from the ninth century. The career of Flann mac Maíle Michíl, mentioned earlier, demonstrates the actualities. He was a literate professional, a *fer léigind*, at Clonmacnoise; he was also the titular head of one of that church's dependent foundations, Cluain Deochra. Flann was the mediator between two layers, with Clonmacnoise as the primary node.

Flann's unifying role was personal but far from unique. Another example, one which sheds even more light on the intersection of literate professionals with secular society, is Suairlech (†870) of Inan (Johnston 2013: 118). Suairlech's death is commemorated in the early medieval chronicles, and is an example of an entry which originated in the Chronicle of Ireland core. His obit is packed with information, informing the reader that Suairlech was a bishop, the abbot of Clonard, an anchorite and a *doctor*. The latter is a term reserved for experts in Christian doctrine and is frequently associated with the Céli Dé, a loose yet influential group of clerics with common ascetic and liturgical interests (Follett 2006: 5–23; Haggart 2006–2007). In fact, Suairlech's connection with Inan securely aligns him with the Céli Dé, as it appears to have been a small anchoritic settlement tied into the Clonard ecclesiastical federation. Suairlech's ascetic commitments in no way militated against ecclesiastical promotion or political influence. As a bishop and abbot of the major monastery of Clonard he was a leading churchman. This is not mere speculation: Suairlech appears elsewhere in the chronicles. He is noted as being present at a *rígdál*, a type of formal royal conference, held in Armagh in 851. This was between Máel Sechnaill mac Maíle Ruanaid (†862), Uí Néill king of Ireland, Matudán (†857), king of Ulaid, Díarmait (†852) and Fethgna (†874), representing Armagh, and Suairlech, at the head of a midland clerical faction. The entry directly names five individuals, three clerics and two kings. These five were each linked into wider networks, whose impact can be traced in the annals. Moreover, Suairlech attended an even more important royal conference a few years later, in 859. This meeting was so significant that it receives a particularly full account in the chronicles. The entry from the Annals of Ulster is worth citing in full.

Righdhal mathe Erenn oc Raith Aedho meic Bricc im Mael Sechnaill rig Temhra, ⁊ im Fethghna comurba Patraicc, 7 im Suairlech comurba Finnio, ic denum sidha 7 caincomraicc fer nErenn, conid asin dail-sin du-rat Cerball, ri Osraighi, oghreir samtha Patraic 7 a comurba, 7 conidh and do-dechaidh Osraighi i ndilsi fri Leth Cuinn, 7 ad-rogaidh Mael Gualai, ri Muman, a dilsi.

A royal gathering of the nobles of Ireland at Rahugh, including Máel Sechnaill the king of Tara, Fethgna successor of Patrick, and Suairlech successor of Finnén, to make peace between the men of Ireland, and as a result of that gathering Cerball, the king of Osraige, gave his full dues to Patrick's congregation and to his successor, and Osraige was alienated to the Northern Half, and Máel Gualai, king of Munster, invoked sureties to guarantee the alienation.

The chronicler uses legalistic language to describe a major redrawing of Ireland's political map in favour of the Uí Néill. Osraige, a powerful regional kingdom in Munster, was removed from that province's sphere into that of Uí Néill, an action which suited Máel Sechnaill and Cerball (†888), king of Osraige (Downham 2004). It concomitantly weakened their Munster rival. The entry encapsulates networks in operation. Suairlech is a central node, just as he was in 851. In fact, three of the protagonists are the same: Máel Sechnaill, Suairlech and Fethgna. Díarmait and Matudán had died in the interim. It is surely telling that Máel Sechnaill's ambitions of suzerainty were underpinned by the support of a clerical party from his own powerbase as well as by the Church of Armagh. The practice of power was dependent on the common interests of ecclesiastical and secular leaders. Suairlech's own role is fascinatingly complex. As an anchorite he was plugged into the Céli Dé network; as a *doctor* he was part of the literate elite; as an ecclesiastical politician he worked closely with kings. Suairlech's network consists of several named individuals and they, in their turn, were connected with yet other identifiable figures. Furthermore, his ecclesiastical leadership was reinforced by the dominant institutional position acquired by centres such as Clonard. The complexity of interests swirling through Suairlech counter-balances the siren schematisation of *Bretha Nemed*. The application of network analysis would surely further enhance an appreciation of these dynamic systems, systems through which learning, power and patronage flowed.

Literate Networks in Literary Sources

The final part of this chapter examines the depiction of literate networks in a different type of source, one which relates to the chronicle material, but at an angle; it also situates these networks within a broader cultural context, widening the focus to include material from the seventh century reinterpreted in the light of the ninth and tenth centuries. This is the context delineated in the vernacular narrative literature. Generally speaking, many early medieval Irish narratives share a legendary pre-Christian past, frequently extending back to the time of Christ and even further (Mac Cana 1980; Ní Bhrolcháin 2009). But this is not always the case. For example, there is a subset of tales where Patrick and other saints, associated with the earliest horizon of christianisation, play major roles (Johnston 2001; Dooley 2004). Some of these narratives, as Joseph Nagy has shown, are deeply concerned with the relationship between the church, secular society and traditional culture (Nagy 1997). Yet others, which are set in the post-conversion era, feature protagonists who are commemorated in the medieval chronicles by their contemporaries. This is a crucial point: such stories, unlike those involving Patrick, fictionalize individuals who died after the practice of chronicle writing had begun. They belong, in a sense, to a definite historical past, rather than to that more distant time which essentially functioned as legend. Due to their often fragmentary nature, however, they have not received the level of attention accorded to those which celebrate great legendary heroes such as Cú Chulainn and Finn mac Cumaill. For instance, the *Táin*, a central text in the Ulster Cycle, has already been examined using the mathematical tools of network analysis (MacCarron and Kenna 2014). Unlike the *Táin*, which is a long and involved tale with a large cast, these narratives are simpler and have far fewer characters. Because of their small scale it is arguably best to consider them as falling into broad groups which are defined by their shared protagonists, expanding the dataset. Most of them consist of short anecdotes such as those concerning the saint and abbot Mo Ling (†697), found in the Book of Leinster (Best and O'Brien 1967: 1236–1242). Many are difficult to date securely and appear to have undergone revision at various points. A good example are those stories centring on the historical king of Connacht, Guaire Aidne (†663) and his relationship with the legendary poet Senchán Torpéist (Ó Coileáin 1974). This narrative type contrasts with the medieval chronicles in several important ways. Three, in particular, are worth highlighting: the stories are retrospective, reflexive and reflective. In other words, they are written long after the deaths of their historical protagonists; the social relationships of the tales are shaped by the imagined past of the protagonists and by the present of the writers, past and present looping into each other; finally, these stories, like many others, idealize the proper shape of society, similar to the schematisation found in the legal material. In contrast, the dominant mode of early medieval chronicle writing is reportage. However, this reportage was determined by the hierarchical frame of Irish society, with all its biases, the same frame informing the tales. Therefore, despite genre contrasts, it is useful to consider whether the actual networks of real people, who are recorded in the chronicles, match those of their fictional alter-egos. The group of stories surrounding Cuimmíne Fota (†662), a literate scholar, are particularly apt for this purpose.

The Cuimmíne of history was commemorated upon his death in 662. His obit in the Annals of Ulster, typical of this early stage of annalistic writing, is short, simply noting that Cuimmíne Fota was a *sapiens*, a scriptural scholar. Fortunately, other sources, including the genealogies, provide additional information. Cuimmíne was born into the west Munster dynasty, Eóganacht Locha Léin. He became abbot of Clonfert, an influential position within the Irish Church (Charles-Edwards 2004: http://www.oxforddnb.com/view/article/51012), making him a significant node in the island's ecclesiatical network. However, things become more complicated once attention is turned from Cuimmíne to the works that he may have written. The reason lies in his name: Cuimmíne with its many variants, including Cumméne and Cummianus, was not uncommon. Historical examples include Cumméne Find (†669) of Iona and the eighth-century Cummian of Bobbio. There is a danger of conflating different individuals who happen to share the same name. Furthermore, Cuimmíne is sometimes confused with another Irish name, Caimín, in verncacular sources (Ó Coileáin 1974: 92). However, it is widely accepted that Cuimmíne Fota wrote the *Penitential of Cummean*, a major Hiberno-Latin text (Bieler 1963: 108–35; Ó Cróinín 1989: 271, 275). Moreover, Dáibhí Ó Cróinín has persuasively supported his authorship of the exegetical *De Figuris Apostolorum* (Ó Cróinín 1989). 'Celebra Iuda', a hymn in honour of the apostles, has also been ascribed to him (Kenney 1929: 266 no. 93). He was clearly a writer of some importance. Scholars have further debated whether Cuimmíne Fota is the same as the Cummian who wrote the Paschal Letter of 632, a crucial source for tracing the Easter Controversy in Ireland (Walsh and Ó Cróinín 1988). While some doubt remains (Lapidge and Sharpe 1985: 78–80), a tentative consensus accepts the identity between Cuimmíne Fota and Cummian (Breen 2009: http://dib.cambridge.org/viewReadPage.do?articleId= a2294). This identity serves to expand Cuimmíne's network further. The Paschal Letter is addressed to Ségéne (†652), abbot of Iona and the otherwise unknown Beccán the hermit. Moreover, the Letter arose out of a major synod of the Irish Church, held at Mag Léne, near Durrow, around 630, placing Cuimmíne at the epicentre of an extensive ecclesiastical network. Indeed, the historical Cuimmíne Fota, even if he is not the same as Cummian, is a significant figure. As an author and high-status churchman he was someone worthy of commemoration by the chroniclers.

The literary Cuimmíne shares some of these characteristics; he is also the subject of a well-developed fictive biography. This Cuimmíne is depicted as being a child of incest and a man whose life colourfully intersected with kings and poets and fools (Ó Coileáin 1974: 92–95; Clancy 1993: 113–117). The dossier of texts in which he features is extensive, including a late Irish Life, a saga, several anecdotes of varying dates, genealogical quatrains and prose fragments (O'Keeffe 1911; Meyer 1902; Kenney 1929: 420–421; Ó Coileáin 1974: 92–101). The genealogical and fragmentary materials have been surveyed by Seán Ó Coileáin, who has argued that they originally formed part of a now incomplete cycle of West Munster tales championing the interests of Cuimmíne's family, Eóganacht Locha Léin (Ó Coileáin 1974). Whether a formal cycle of tales existed or not, there can be no doubt that the extant texts operate within a shared legendary history and through a common cast of stereotyped characters. The earliest narrative stratum seems to be ninth and tenth century but it continued to evolve long after, becoming ever more complex.

This complexity is as much one of relationships between characters as of redaction. The main protagonists are Cuimmíne, the *oínmit* 'fool' Comgán Mac Dá Cherda, Guaire, king of Connacht and, more tangentially, St Íte (†570/577), founder of Killeedy. Cuimmíne, Guaire and Íte are all genuine historical figures. What of Mac Dá Cherda? The death of a Comgán Mac Dá Cherda is recorded in *Chronicum Scotorum* and the Annals of Inisfallen, at 641 and 645 respectively. It has been speculated that he is the Comgán, abbot of Emly who appears elsewhere as one of the ecclesiastical guarantors of the so-called West Munster Synod (Byrne 1973: 216–217). If this is the case, he has certainly undergone a remarkable transformation; the Holy Fool of the stories bears no trace of Comgán's ecclesiastical career and the identification seems unlikely. In any case, Cuimmíne is closely linked to the other three. He is imagined to be the uterine brother of Guaire and the foster-brother, sometimes uterine sibling, of Mac Dá Cherda. Íte, adopting her traditional aspect as foster-mother of the Irish saints (Johnston 2000), rescues Cuimmíne as an infant and raises him. It is noteworthy that the historical horizon of the narratives is generally consistent. Guaire, Mac Dá Cherda and Cuimmíne Fota were all contemporaries. Íte is the one outlier: her death in 570 or 577 makes her an exceptionally old foster-mother to a man who died in 662, although it is just about possible. Significantly, however, Íte is a central figure in the idealized networks connecting sixth- and seventh-century Irish saints. It seems obvious that she fulfils the maternal node which was a feature of actual kinship networks.

At this point it is worth asking if there are any commonalities between the fictional Cuimmíne and his historical inspiration. Indeed, are there any signs of this historical inspiration beyond his name and genealogical connections? One tale, in particular, is worth investigating. It has a coherent narrative trajectory and envisages the interaction of secular and ecclesiastical networks. Furthermore, a full appreciation of its dynamics requires a knowledge of the wider tradition into which it fits. This is the Old Irish saga *Comrac Líadaine ocus Cuirithir*. The text tells of the love between two poets and their choice of sexual renunciation, under the direction of Cuimmíne Fota, over consummation. Even Cuimmíne's guidance is not enough; Cuirithir is obliged to go into penitential exile to avoid Líadain's sexual temptations, at first in Ireland and then overseas. This invokes the well-known Irish practice of *peregrinatio* (Charles-Edwards 1976; Johnston 2016), normally associated with exile clerics such as Columbanus (†615). The tale ends with Líadain losing Cuirithir but winning her soul's salvation (Meyer 1902: 26–27). It is tempting to see Cuimmíne's portrayal as directly echoing his real reputation as an authority on penance. The same reputation appears to underlie the tradition that he gave the veil to Digde, the supposed author of the Old-Irish poem *Caillech Bérre*, a notably penitential production (Ó Coileáin 1974: 108–109). Mac Dá Cherda, who is described as being simultaneously chief-*fili* and fool of Ireland, plays a complementary part. Ostensibly helping Cuirithir, it is his verses which inspire the two poets to enter the religious life and seek out Cuimmíne (Meyer 1902: 12–18). Thomas Clancy, in a convincing analysis of the tale, highlights Mac Dá Cherda's actions as the catalyst which pushes the poets from the secular to the religious (Clancy 1993: 120–122).

The saga is a story of networks, their fluctuation, creation and destruction. *Comrac Líadaine ocus Cuirithir* presents its audience with two simple networks. These are explicitly linked through Líadain and Cuirithir and implicitly through Cuimmíne's unstated, but well-known, relationship with Mac Dá Cherda. The tale opens in a world of poets, including the two would-be lovers, their retinues and Mac Dá Cherda. This is replaced, through Mac Dá Cherda's actions, with an ecclesiastical network consisting of Cuimmíne, Líadain and Cuirithir. This network too breaks apart: Líadain and Cuirithir only find salvation in isolation. However, the implicit network, the one which joins Mac Dá Cherda with Cuimmíne Fota, is what drives the action of the narrative. Their status as siblings or foster-brothers is never mentioned in the saga, but there was no need; it is the bedrock of their interactions in several stories. In effect, Mac Dá Cherda, acting as a proxy for Cuimmíne, prevents rather than enables the physical fulfilment of the poets' relationship. This may seem far removed from actual history but, as in the case of Urard mac Coise and Flann mac Maíle Michíl, the tale presents a network joining clerics and poets. Furthermore, when *Comrac Líadaine ocus Cuirithir* is situated within its wider story complex, the network expands to include the king of Connacht, the legendary poet Senchán

Torpéist, and the churches of Clonfert and Ardfert (O'Keeffe 1911). Without a doubt, the Cuimmíne of these stories is not the Cuimmíne of history, although shadows of the original flit through the surviving texts. What he is, instead, is an individualized social myth. The fictional Cuimmíne and his connections form a template, the template of what the career of an ecclesiastical literate professional might resemble. He is the confidant of kings and of poets; he is equally at home in penitential austerity or at the royal court. The ideal trio of these stories consists of a poet, a king and a cleric, mirroring, in part, the scheme of the law tracts. However, this template is filled with names familiar from history, rounded out through characters and through plots which are driven by social expectations. This combination of factors means that the network of the tales does resemble that of the medieval chroniclers, although it also looks towards the society imagined in the vernacular laws. It is a glimpse of how the literate elite imagined their world. These narratives inhabit the space between the quantitative realities of the chronicles and the ideologies of law.

The Potentials of Network Theory

As this paper has shown, the world of the literate elite, their networks and connections, is richly represented in the extant texts. These come in a variety of genres and scales, ranging from medieval chronicles that commemorate thousands of individuals to short tales which name a handful. The focus has been largely on the medieval chronicles. Their sheer size makes them the best corpus of material for any quantitative analysis, including network theory. Moreover, they can be cross-referenced with other sources such as genealogies, fragmentary anecdotes, sagas and law tracts. The medieval chronicles provide a spine of contemporary reportage which documents the interactions of socio-political networks. However, it is useful to deepen contextualization by including other types of text, ones which provide reflective and reflexive insights into the nature of early medieval Irish society. The case study of Cuimmíne Fota is one example of how this might be done. Conceptually, network theory will give scholars a new tool for approaching these texts, allowing for an understanding of how the society as a whole functioned and how it perceived itself to function. Previously, this has been analyzed through an emphasis on the history of institutions with the result that individuals, their actions, and their links with other individuals, have been under-appreciated. By considering one definable group in a specific time frame, this paper has suggested a model of how network theory can be usefully employed, one which offers a fresh way of approaching the early medieval Irish past and the people who lived it and created it.

Bibliography

The Irish Chronicles

Charles-Edwards, T. M. (Ed.). (2006). *The chronicle of Ireland* (Vol. 2). Liverpool: Liverpool University Press.

Hennessy, W. M. (Ed. & Trans.) (1866). *Chronicum Scotorum: A chronicle of Irish affairs form the earliest times to AD 1135, with a supplement containing events from 1141–1156* (The Rolls Series 46). London: Longmans.

Mac Airt, S., & Mac Niocaill, G. (Eds. & Trans.) (1983). *The Annals of Ulster (to A.D. 1131)*. Dublin: Dublin Institute for Advanced Studies.

Mac Airt, S. (Ed. & Trans.) (1951). *The Annals of Inisfallen*. Dublin: Dublin Institute for Advanced Studies.

Murphy, D. (Ed.). (1896). *The Annals of Clonmacnoise, being annals of Ireland from the earliest period to A.D. 1408, translated into English, A.D. 1627, by Conell Mageoghagan*. Dublin: University Press, Royal Society of Antiquaries of Ireland.

O'Donovan, J. (Ed. & Trans.) (1848–1851). *Annála Rioghachta Éireann: Annals of the Kingdom of Ireland by the Four Masters, from the earliest period to the year 1616* (Vol. 7). Dublin: University of California Press.

Radner, J. N. (Ed. & Trans.) (1978). *Fragmentary Annals of Ireland*. Dublin: Dublin Institute for Advanced Studies.

Stokes, W. (Ed.) (1895–1897). The Annals of Tigernach. *Revue Celtique* 16: 374–419; 17: 6–33, 119–263, 337–420; 18: 9–59, 150–197, 267–303.

Other Primary Sources

Best, R. I., & O'Brien, M. A. (Eds.). (1967). *The Book of Leinster, formerly Lebar na Núachong-bála* (Vol. 5). Dublin: Dublin Institute for Advanced Studies.

Bieler, L. (Ed. & Trans.) (1963). *The Irish penitentials*. Dublin: Dublin Institute for Advanced Studies.

Byrne, M. E. (Ed.) (1908). Airec Menman Uraird maic Coise. In O. Bergin et al. (Eds.), *Anecdota from Irish manuscripts* (Vol. 2, pp. 42–76). Halle and Dublin: Max Niemeyer and Hodges Figgis.

Gwynn, E. J. (Ed.) (1940–1942). An Old-Irish tract on the privileges and responsibilities of poets. *Ériu* 13: 1–60, 220–236.

Meyer, K. (Ed. & Trans.) (1902). *Liadain and Curithir: An Irish love-story of the ninth century*. London: D. Nutt.

O'Brien, M. A. (Ed.). (1962). *Corpus genealogiarum Hiberniae* (Vol. 1). Dublin: Dublin Institute for Advanced Studies.

O'Keeffe, J. G. (Ed. & Trans.) (1911). Mac Dá Cherda and Cummaine Fota. *Ériu* 5: 18–44.

Walsh, M., & Ó Cróinín, D. (Eds. & Trans.) (1988). *Cummian's letter* De Controversia Paschali, *together with a related Computistical Tract* De Ratione Computandi. Toronto: Pontifical Institute of Mediaeval Studies.

Secondary Sources

Bitel, L. (1996). *Land of women: Tales of sex and gender in early Ireland.* Ithaca, NY: Cornell University Press.

Breatnach, L. (2011). *The early Irish law text Senchas Már and the question of its date.* Cambridge: Department of Anglo-Saxon, Norse and Celtic. E. C. Quiggin Memorial Lectures 12.

Breen, A. (2009). Cummian (Cummíne, Cumméne) Foto. In J. McGuire & J. Quinn (Eds.), *Dictionary of Irish biography.* Cambridge: Cambridge University Press.

Byrne, F. J. (1973). *Irish kings and high-kings.* London: Batsford.

Charles-Edwards, T. M. (1976). The social background to Irish *peregrinatio. Celtica, 11,* 43–59.

Charles-Edwards, T. M. (1993). *Early Irish and Welsh kinship.* Oxford: Oxford University Press.

Charles-Edwards, T. M. (2000). *Early Christian Ireland.* Cambridge: Cambridge University Press.

Charles-Edwards, T. M. (2004). Connacht, saints of (*act. c.*400–*c.*800). In L. Goldman (Ed.), *Oxford dictionary of national biography.* Oxford: Oxford University Press. Accessed April 27, 2015 from http://www.oxforddnb.com/view/article/51012

Clancy, T. O. (1993). Fools and adultery in some early Irish texts. *Ériu, 44,* 105–124.

Connon, A. (2000). The *Banshenchas* and the Uí Néill Queens of Tara. In A. Smyth (Ed.), *Seanchas: Studies in early and medieval Irish archaeology, history, and literature in honour of Francis J. Byrne* (pp. 98–108). Dublin: Four Courts Press.

Cunningham, B. (2010). *The Annals of the Four Masters: History, kingship and society in the early seventeenth century.* Dublin: Four Courts Press.

Doherty, C. (2005). Kingship in early Ireland. In E. Bhreathnach (Ed.), *The kingship and landscape of Tara* (pp. 3–31). Dublin: Four Courts Press.

Dooley, A. (2004). The date and purpose of *Acallam na Senórach. Éigse, 34,* 97–126.

Downham, C. (2004). The career of Cearbhall of Osraighe. *Ossory, Laois and Leinster, 1,* 1–18.

Dumville, D. N. (1982). Latin and Irish in the *Annals of Ulster,* A.D. 431–1050. In D. Whitelock, R. McKitterick, & D. Dumville (Eds.), *Ireland in early medieval Europe: Studies in memory of Kathleen Hughes* (pp. 320–341). Cambridge: Cambridge University Press.

Evans, N. (2010). *The present and the past in medieval Irish chronicles.* Woodbridge: Boydell and Brewer.

Flechner, R. (2013). The chronicle of Ireland: Then and now. *Early Medieval Europe, 21*(4), 422–454.

Follett, W. (2006). *Céli Dé in Ireland: Monastic writing and identity in the early middle ages.* Woodbridge: Boydell and Brewer.

Grabowski, K., & Dumville, D. N. (1984). *Chronicles and annals of medieval Ireland and Wales: The Clonmacnoise-group texts.* Woodbridge: Boydell and Brewer.

Haggart, C. (2006–2007). The *Céli Dé* and the early medieval Irish church: A reassessment. *Studia Hibernica* 34: 17–62.

Hughes, K. (1972). *Early Christian Ireland: An introduction to the sources.* Cambridge: Cambridge University Press.

Ireland, C. (1996). Aldfrith of Northumbria and the learning of a *sapiens.* In K. Klar, E. Sweetser, & C. Thomas (Eds.), *A Celtic florilegium: Studies in memory of Brendan O'Hehir* (pp. 63–77). Andover MA: Celtic Studies Publications.

Johnston, E. (2000). Íte: Patron of her people? *Peritia, 14,* 421–428.

Johnston, E. (2001). The salvation of the individual and the salvation of society in *Siaburcharpat Con Culaind. CSANA Yearbook, 1,* 109–25.

Johnston, E. (2013). *Literacy and identity in early medieval Ireland.* Woodbridge: Boydell and Brewer.

Johnston, E. (2016). Exiles from the edge? The Irish contexts of peregrinatio. In R. Flechner & S. Meeder (Eds.), *The Irish in early medieval Europe: Identity, culture and religion* (pp. 38–52). Basingstoke: Palgrave Macmillan.

Kehnel, A. (1995). *Clonmacnois—The church and lands of St Ciarán, change and continuity in an Irish monastic foundation (6th to 16th century).* Münster: LIT Verlag.

Kelly, F. (1988). *A guide to early Irish law*. Dublin: Dublin Institute for Advanced Studies.
Kenney, J. F. (1929). *The Sources for the early history of Ireland: Ecclesiastical: An introduction and guide*. Columbia: Columbia University Press.
Lapidge, M., & Sharpe, R. (1985). *A bibliography of Celtic-Latin literature 400–1200*. Dublin: Royal Irish Academy.
Mac Cana, P. (1980). *The learned tales of medieval Ireland*. Dublin: Dublin Institute for Advanced Studies.
Mac Carron, P., & Kenna, R. (2012). Universal properties of mythological networks. *EPL* 99. doi: 10.1209/0295-5075/99/28002
MacCarron, P., & Kenna, R. (2014). Network analysis of *Beowulf*, the *Iliad* and the *Táin Bó Cúailnge*. In K. Antoni & D. Weiss (Eds.), *Sources of mythology: Ancient and contemporary myths*. Proceedings of the Seventh Annual International Conference on Comparative Mythology (15–17 May 2013, Tübingen), pp. 125–141. Tübingen: Religionswissenschaft: Forschung und Wissenschaft.
Mac Niocaill, G. (1975). *The medieval Irish annals*. Dublin: Dublin Historical Association.
MacCotter, P. (2008). *Medieval Ireland: Territorial, political and economic divisions*. Dublin: Four Courts Press.
Mc Carthy, D. (1998). The chronology of the Irish annals. *Proceeding of the Royal Irish Academy (C), 99*, 203–255.
Mc Carthy, D. (2008). *The Irish annals: Their genesis, evolution and history*. Dublin: Four Courts Press.
Mc Carthy, D., & Breen, A. (1997). An evaluation of astronomical observations in the Irish annals. *Vistas in Astronomy, 41*, 117–138.
Mc Cormick, M., et al. (2007). Volcanoes and the climate forcing of Carolingian Europe, A.D. 750–950. *Speculum, 82*(4), 865–895.
McGowan, M. K. (2004). The Four Masters and the governance of Ireland. *The Journal of Celtic Studies, 4*, 1–41.
Nagy, J. F. (1997). *Conversing with angels and ancients: Literary myths of medieval Ireland*. Ithaca, NY: Cornell University Press.
Ní Bhrolcháin, M. (2009). *An introduction to early Irish literature*. Dublin: Four Courts Press.
Ní Dhonnchadha, M. (1982). The guarantor list of Cáin Adomnáin, 697. *Peritia, 1*, 178–215.
Ní Dhonnchadha, M. (2004). Adomnán [St Adomnán] 627/8?–704. In L. Goldman (Ed.), *Oxford dictionary of national biography*. Oxford: Oxford University Press. Accessed April 27, 2015 from http://www.oxforddnb.com/view/article/110
Ó Catháin, S. (1933). Some studies in the development from Middle to Modern Irish, based on the Annals of Ulster. *Zeitschrift für Celtische Philologie, 19*, 1–47.
Ó Coileáin, S. (1974). The structure of a literary cycle. *Ériu, 25*, 88–125.
Ó Corráin, D. (1972). *Ireland before the Normans*. Dublin: Gill and Macmillan.
Ó Corráin, D. (1978). Nationality and kingship in pre-Norman Ireland. In T. W. Moody (Ed.), *Nationality and the pursuit of national independence* (pp. 1–35). Belfast: The Appletree Press.
Ó Corráin, D. (1998). Creating the past: The early Irish genealogical tradition. *Peritia, 12*, 177–208.
Ó Corráin, D., Breatnach, L., & Breen, A. (1984). The laws of the Irish. *Peritia, 3*, 382–438.
Ó Cróinín, D. (1989). Cummianus Longus and the iconography of Christ and the apostles in early Irish literature. In L. Breatnach, K. McCone, & D. Ó Corráin (Eds.), *Sages, saints and storytellers: Celtic studies in honour of Professor James Carney* (pp. 268–279). An Sagart: Maynooth.
ÓFloinn, R. (1998). Clonmacnoise: Art and patronage in the early medieval period. In H. King (Ed.), *Clonmacnoise studies* (Vol. 1, pp. 87–100). Dublin: The Stationary Office.
Radner, J. (1999). Writing history: Early Irish historiography and the significance of form. *Celtica, 23*, 312–325.
Russell, P. (2004). Cormac mac Cuilennáin (*d.* 908). In L. Goldman (Ed.), *Oxford dictionary of national biography*. Oxford: Oxford University Press. Accessed April 27, 2015 from http://www.oxforddnb.com/view/article/6319
Watkins, C. (1976). Sick-maintenance in Indo-European. *Ériu, 27*, 21–25.

How Quantitative Methods Can Shed Light on a Problem of Comparative Mythology: The Myth of the Struggle for Supremacy Between Two Groups of Deities Reconsidered

David Weiß

Abstract This chapter treats a well-known mythic theme from Japanese mythology: the struggle of the 'Earthly Deities' against the 'Heavenly Deities' for government of the world. To this day, many scholars interpret this episode in a purely historical fashion, that is, as a reflection of the conflict between two political groups. However, this approach fails to explain the wide distribution of very similar myths in Eurasia. Examples include the mythologies of Greece (Titans versus Olympians), India (Devas versus Asuras) and Scandinavia (giants versus gods). The historical-comparative approaches of Georges Dumézil and Michael Witzel take the wide distribution of this theme into account and explain it with common origin. Dumézil believes that the Indo-Europeans shared a tripartite ideology which was represented in their social structure as well as in their myths. According to this theory, one group of deities represented the 'functions' of 'king/priest' and of 'warrior' while the other group (which loses the struggle) represents the function of 'cultivator/herder'. Witzel, on the other hand, argues that the theme can be explained as one episode in the 'Laurasian story line' which was created by our ancestors some 40,000 years ago in or near south-western Asia and subsequently diffused parallel to the spread of the human race. After a discussion of these two theories, I will suggest how quantitative approaches like social network analysis, phylogenetics or principal component analysis might enable us to counter-check these hypotheses and be instrumental in either refuting them or placing them—and thereby the often highly speculative field of comparative mythology as such—on a firmer scientific fundament.

D. Weiß (✉)
Department of Japanese Studies, Institute of Asian and Oriental Studies, Eberhard Karls Universität Tübingen, Wilhelmstr. 90, 72074 Tübingen, Germany
e-mail: david.weiss@uni-tuebingen.de

© Springer International Publishing Switzerland 2017
R. Kenna et al. (eds.), *Maths Meets Myths: Quantitative Approaches to Ancient Narratives*, Understanding Complex Systems,
DOI 10.1007/978-3-319-39445-9_12

On the Definition of "Myth"

Since the term "myth" has been, and continues to be, used in various ways by different scholars, it might be in order to start with some remarks on its definition. Since the days of Jakob Grimm (1785–1863) and Wilhelm Grimm (1786–1859), who collected traditional tales in Germany and are often regarded as the founders of the discipline of folklore, folklorists have distinguished myths from folktales and legends. A representative treatment of this distinction is offered by William Bascom, who categorizes all three genres under the rubric of "prose narratives" but claims that they can be distinguished based on whether they are believed to be true and according to their temporal and spatial setting:

- *Folktales are prose narratives which are regarded as fiction.* [...] [They] may be set in any time and any place, and in this sense they are almost timeless and placeless. [...]
- *Myths are prose narratives which, in the society in which they are told, are considered to be truthful accounts of what happened in the remote past.* [...]
- *Legends are prose narratives which, like myths, are regarded as true by the narrator and his audience, but they are set in a period considered less remote, when the world was much as it is today* (Bascom 1984, 8–9, emphasis in the original).

These clear-cut definitions are certainly useful for a categorization of different tales. However, it has been pointed out that prior to the Grimms' pioneering studies "neither English nor any other European language traditionally made this distinction [...]. Given the historical contingency and, indeed, the very short life span even in European tongues of this tripartite classification of oral traditional narrative genres, it would be remarkable to find that this system of classification is otherwise a human universal." (Csapo 2005, 6) To do justice to Bascom, he explicitly states that "Myth, legend, and folktale are not proposed as universally recognized categories but as analytical concepts which can be meaningfully applied cross-culturally even when other systems of 'native categories' are locally recognized" (Bascom 1984, 10).

A more serious problem of the distinction between myth, legend and folktale is acknowledged by Bascom himself:

> "In passing from one society to another through diffusion, a myth or legend may be accepted without being believed, thus becoming a folktale in the borrowing society, and the reverse may also happen. It is entirely possible that the same tale type may be a folktale in one society, a legend in a second society, and a myth in a third. Furthermore, in the course of time fewer and fewer members of a society may believe in a myth, and especially in a period of rapid cultural change an entire belief system and its mythology can be discredited. Even in cultural isolation, there may be some skeptics who do not accept the traditional system of belief" (Bascom 1984, 13).

For this reason the lines between myth, legend and folktale should not be drawn too sharply in a comparative intercultural study of similar narrative motifs like the one attempted in this chapter. A more complex definition, which better fits the purpose of the present study, was suggested by Doty (1986, 11):

> A mythological corpus consists of (1) a usually complex network of myths that are (2) culturally important (3) imaginal (4) stories, conveying by means of (5) metaphoric and symbolic diction, (6) graphic imagery, and (7) emotional conviction and participation, (8) the primal, foundational accounts (9) of aspects of the real, experienced world and (10) humankind's roles and relative statuses within it.
>
> Mythologies may (11) convey the political and moral values of a culture and (12) provide systems of interpreting (13) individual experience within a universal perspective, which may include (14) the intervention of suprahuman entities as well as (15) aspects of the natural and cultural orders. Myths may be enacted or reflected in (16) rituals, ceremonies, and dramas, and (17) they may provide materials for secondary elaboration, the constituent mythemes having become merely images or reference points for a subsequent story, such as a folktale, historical legend, novella, or prophecy.

It is important to note that a particular tale does not need to meet all the 17 criteria listed in this definition to be regarded as a myth. Ludwig Wittgenstein's concept of "family resemblances" offers a useful model to describe the relationship between particular myths (cf. Csapo 2005, 8). Wittgenstein explained this concept by using the example of "games":

> "66. Consider for example the proceedings that we call 'games'. I mean board-games, card-games, ball-games, Olympic games, and so on. What is common to them all? [. . .] [Y]ou will not see something that is common to *all*, but similarities, relationships, and a whole series of them at that. [. . .] Look for example at board-games, with their multifarious relationships. Now pass to card-games; here you find many correspondences with the first group, but many common features drop out, and others appear. When we pass next to the ball-games, much that is common is retained, but much is lost—Are they all 'amusing'? [. . .] Or is there always winning and losing, or competition between players? [. . .] In ball-games there is winning and losing; but when a child throws his ball at the wall and catches it again, this feature has disappeared. [. . .] And the result of this examination is: we see a complicated network of similarities overlapping and criss-crossing: sometimes overall similarities, sometimes similarities of detail.
>
> 67. I can think of no better expression to characterize these similarities than 'family resemblances'; for the various resemblances between members of a family: build, features, colour of eyes, gait, temperament, etc. etc. overlap and criss-cross in the same way—And I shall say: 'games' form a family" (Wittgenstein 1953, 31–32, emphasis in the original).

Myths can be understood as a family in Wittgenstein's sense as well, in which case Doty's 17 criteria would represent a catalogue of some of the most important resemblances between particular members of this family. A point which, perhaps, calls for elaboration is the second on Doty's list: myths are culturally important. "What makes a story a myth", according to Csapo (2005, 134), "is the fact that it is received by a given society and that a given society participates in its transmission. [. . .] [In this process] the intentions of a given society in transmitting a narrative may have nothing to do with the purposes of its author."

The Sources of Japanese Mythology

The main sources of Japanese mythology are the *Kojiki* ("Record of Ancient Matters")[1] and the *Nihon shoki* ("Chronicles of Japan")[2]. Both of these chronicles were compiled by imperial order at the beginning of the eighth century. They basically deal with the same subject matter: a mythical part which relates the origin of gods and the Japanese islands is followed by a legendary part which describes the heroic conquest of these islands by the first emperors and, finally, a more thoroughly historical account of the reigns of later emperors. These sections seamlessly blend into each other and the unbroken dynasty of the imperial family which stretches back into the divine age is the focus of the narrative. One major difference between the two works, however, is that the *Nihon shoki* is more heavily influenced by (mainly Confucianist) ideas from China. For this reason, the *Nihon shoki* is more rationalistic with regard to the myths, providing numerous variants of particular mythic episodes which are often mutually contradictory. In this chapter, I will focus on the *Kojiki* which gives a more consistent account of the Japanese divine age and therefore probably lends itself better to quantitative approaches.

Summary of the Struggle Between Heavenly and Earthly Deities

In this section, I provide a brief summary of the *Kojiki's* account of the divine age with special focus on the struggle between the Heavenly and the Earthly Deities. At the beginning, the primordial parents Izanagi and Izanami created the Japanese islands as well as various deities by sexual intercourse. However, when Izanami gave birth to the fire god, she was burned and died. Her husband Izanagi tried unsuccessfully to retrieve her from the netherworld. After this failed attempt, he ritually cleaned himself of the filth of the netherworld in the mouth of a river. During this purification various deities came into existence. The most important of these were the sun goddess Amaterasu, who came into existence when he washed his left eye, the moon god Tsukiyomi, who came into existence when he washed his right eye, and, lastly, the god Susanowo, who emerged when Izanagi washed his nose. To these three deities Izanagi entrusted the government of the world: Amaterasu was assigned to rule the Plain of High Heaven, while the Land of Night was apportioned to Tsukiyomi and the Plain of the Ocean to Susanowo. Whereas the sun and moon deities ruled their respective realms as Izanagi had ordered, Susanowo cried all the

---

[1] For English translations, see Chamberlain (1982 [1882]) and Philippi (1968); for a recent German translation with extensive commentary, see Antoni (2012).

[2] For an English translation, see Aston (1956[1896]); for a German translation of the mythological section, see Florenz (1901).

time, which caused the rivers and oceans to dry out and the mountains to wither. Consequently, Izanagi banished him.

Next, Susanowo ascended to the Plain of High Heaven in order to say farewell to his sister, the sun goddess. However, once inside his sister's realm, Susanowo destroyed her rice fields, desecrated the most holy precinct and killed one of her weaving maidens. Terrified, Amaterasu retreated into a rock cave, shrouding the world in darkness. With some effort the other deities managed to lure her out again. A fine was imposed on Susanowo and he was banished for a second time. He descended to the land of Izumo by the Sea of Japan (East Sea), where he met an old couple and their daughter. The father introduced himself as the son of an Earthly Deity and explained that he and his wife had originally had eight daughters, but every year a giant serpent had come and devoured one. Now only the last daughter was left and the serpent would soon come to get her too. Susanowo killed the serpent by trickery and married the girl. As their descendant in the sixth generation, Ohokuninushi was born.

Ohokuninushi is usually regarded as the paramount of the Earthly Deities. The next section in the mythical plot deals with his deeds. Ohokuninushi's 80 brothers wanted to marry a princess of Inaba, a country close to Izumo. Ohokuninushi accompanied them as their servant, carrying their luggage. The princess, however, chose Ohokuninushi as her husband. Therefore, his enraged brothers pursued and killed him. When he was miraculously resurrected, they killed him again. Ohokuninushi was resurrected a second time and fled to the Land of Roots, the netherworld where Susanowo now ruled. There he fell in love with Susanowo's daughter and married her. He had to pass some deadly tests to prove his worthiness to Susanowo, however. With the help of his new wife, he succeeded and managed to escape from the Land of Roots while Susanowo was asleep, stealing the latter's daughter as well as his magical weapons. Susanowo woke up and pursued him to the border of his realm. From there he shouted after Ohokuninushi, commanding him to take his daughter as his main wife and use the weapons to slay his brothers and become the ruler of the land. Ohokuninushi followed this advice and built a palace in Izumo. The next section deals with his many love affairs and the many children born to him by different wives. Then, Ohokuninushi, with the help of some other deities, formed and solidified the land.

In the meantime, the sun goddess Amaterasu decreed that her son should rule the Central Land of Reed Plains (i.e. Japan or, perhaps more correctly, the earth). However, when her son looked down from the Plain of High Heaven, he saw that the earth was inhabited by a great number of wild Earthly Deities. Consequently, some deities were sent in order to subdue the Earthly Deities, but they went over to the enemy. Finally, Amaterasu sent the two deities Takemikazuchi and Ame no Torifune. They descended to Izumo and asked Ohokuninushi if he would yield the authority over the Central Land of Reed Plains to Amaterasu's son. He replied that they should talk to his two sons first. After a demonstration of Takemikazuchi's superior strength, the sons complied with Amaterasu's wishes. Thus, Ohokuninushi yielded his land to the Heavenly Deities on the condition that he may live in a palace as great as that of Amaterasu's descendants.

Takemikazuchi returned to the Plain of High Heaven and reported the subjugation of the Central Land of Reed Plains. However, Amaterasu's son insisted that *his* son, Ninigi, should be sent to rule over the land in his stead. Thus Ninigi descended in the company of the ancestor gods of the five leading families of court nobility, carrying the imperial regalia. Interestingly, he descended not to Izumo but to Tsukushi in Kyūshū, from where his grandson Iharebiko, the legendary first emperor of Japan, conquered the Japanese islands.

Historical Interpretations

The most common interpretation of the struggle between Earthly and Heavenly Deities in Japanese myth holds that it reflects historical events. As we have seen, the setting of the myths dealing with Earthly Deities is always Izumo (or the neighbouring Inaba). From archaeological findings we know that Izumo was an important political and cultural centre until the later sixth century, when it is thought to have been incorporated into the emerging Yamato central state (Piggott 1989). Most Japanese scholars agree that the myth of the cession of the land to the Heavenly Deities reflects exactly this process. The Heavenly Deities, they claim, are the gods worshipped by the imperial family and the court aristocracy. Amaterasu, the ancestress of the imperial family, holds the supreme position in this pantheon. According to this hypothesis, the Izumo deities, who were originally unrelated to the Yamato pantheon, were incorporated into the central mythology as Earthly Deities, in a clearly inferior—though still prominent—position (cf. Matsumura 1951, 1954–1958, vol. 3, 442–443, 486–499; Oka 2012, 96–97; Takagi 1973, 248–251). According to Joan Piggott (1989, 62), the myth of the cession of the land "sacralised the relationship between Izumo and Yamato. What was historically a political compromise was now portrayed as a timeless and solemn agreement between Ohonamuchi [= Ohokuninushi] and Amaterasu's descendants. […] The myth sanctified Yamato's paramountcy, but it also guaranteed the continuance of Ohonamuchi's cult and the power of its priest-rulers". In this way, Piggott continues, the retelling of the Izumo myths in the Yamato chronicles "skilfully legitimated the political status quo in the early eighth century: Izumo deities such as Ohonamuchi and Susano-o were appropriately subordinated to Yamato's cult deities; Izumo was described as 'the land of darkness' […] in contrast to Amaterasu's sunny 'high heavenly plain' […]; and Ohonamuchi, lord of the land of Izumo, was portrayed as ceding his divine power over Izumo to Amaterasu's descendants" (Piggott 1989, 67–68).

While it is not always clear to which group a specific deity belongs, the most ambiguous case is presented by Susanowo: he is a brother of Amaterasu and should thus be counted among the Heavenly Deities; however, he is also the protagonist of many tales set in Izumo. He marries an Earthly Deity and is the ancestor of Ohokuninushi, the paramount Earthly Deity, whom he commands to rule over the land and whom he—if involuntarily—provides with the weapons which are necessary for this task. His role in the mythology of the Earthly Deities is thus parallel to the one played by Amaterasu in the mythology of the Heavenly Deities. It has been suggested that Susanowo was invented by the myth-writers at the Yamato court in order to link the two mythical traditions which were originally independent (Matsumura 1954–1958, vol. 2, 605; Tsuda 1963, 576–595).

Parallels in Mythologies Outside of Japan

However, historical interpretations fail to explain the numerous examples of a comparable struggle between two groups of deities which can be observed in many mythologies all over the world. These include, among others, the Devas and Asuras in India, the old and the new gods in Mesopotamia, the Titans and the Olympians in Greece as well as the gods and the giants in Scandinavian myth (cf. van Baaren 1984, 11). In this section, I will briefly discuss the Scandinavian and Greek cases in order to explicate some of the parallels to Japanese mythology. The discussion will be based mainly on the *Prose Edda* by Snorri Sturluson (1179–1241) and Hesiod's *Theogony* (eighth century BC).

In the Greek and Scandinavian cases—as well as in Japanese mythology—the two groups of deities are related by blood. The Titans Kronos and Rhea are the parents of the Olympians Hestia, Demeter, Hera, Hades, Poseidon and Zeus (Hard 2004, 67). In Scandinavian mythology, "Odin was the son of the giantess Bestla, Tyr of the giant Hymir, and Loki of the giant Farbauti. Heimdal was the son of the nine daughters of Aegir, who is sometimes considered as a god and sometimes as a giant" (Oosten 1984, 198; cf. de Vries (1956–1957), vol. 1, 244). As in Japan, the distinction between the two groups is not always clear in Scandinavian myth.

Since Kronos was warned that he would be overpowered by his own son, he swallowed all of his children at birth. However, when Zeus was born, Rhea gave Kronos a stone wrapped in swaddling-clothes to swallow instead of the child. When Zeus was grown up, he forced Kronos to vomit up his siblings and then, together with his siblings, started a war against the Titans, which he won after 10 years. After being defeated by the Olympians, the Titans were imprisoned in the netherworld of Tartaros (Caldwell 1987, 53–57, 63–67). A similar passage can be found in the Japanese myth of the cession of the land. Ohokuninushi here promises that he will conceal himself in the "less-than-one-hundred eighty road-bendings" (Philippi 1968, 134). This strange expression probably refers to a kind of netherworld. In any case it refers to a far removed place (Philippi 1968, 134–35, n. 7) and is thus comparable to Tartators, which, according to the *Theogony,* is situated "as far below the earth as sky is above the earth" (Caldwell 1987, 67).

Another feature of the Japanese myth is more closely mirrored in Scandinavian mythology: there is a mediator who connects the two opposing groups. We have already seen that in Japanese myth this role is played by Susanowo, who seems to belong to neither or both of the groups. In Scandinavian myth, the same role is played by Loki, who has been fittingly called the "arch-mediator [. . .] of Norse mythology" (Haugen 1967, 863). Although he is counted among the gods, his father was a giant and he fathers most of his children by a giantess (Byock 2005, 38–39; cf. Hyde 2010, 97). Interaction between giants and gods virtually always takes place due to Loki's initiative. While the Greek Prometheus shares many characteristics with Susanowo and Loki, he does not occupy a comparable position as mediator between the gods—he rather mediates between gods and humans than between the Titans and the Olympians (Caldwell 1987, 58–61).

A last feature common to all three mythologies is that intermarriage occurs between the two groups of deities. According to the *Theogony*, Zeus married his two aunts, Themis and Mnemosyne, who belonged to the Titans, as well as Leto, a daughter of the Titans Koios and Phoibe (Caldwell 1987, 76–77). In Scandinavian myth, marriages between giants and gods seem to be rather the norm than an exception: the god Njord married the giantess Skadi and Freyr married the giantess Gerd. Furthermore, Odin seduced the giantesses Gunnlod and Rind, but did not marry them, and Loki, as mentioned, fathered three monstrous children by the giantess Angrboda (Byock 2005, 39). The giants, on the other hand, often tried to gain goddesses as wives but never succeeded (Oosten 1985, 38). The same is true for Japanese myth, where a number of Heavenly Deities took Earthly Deities as wives: the first is Susanowo, if he is counted as one of the Heavenly Deities (Philippi 1968, 88–92), another one is Ame no Wakahiko who was sent by the Heavenly Deities to subdue the Earthly Deities but instead married a daughter of Ohokuninushi (Philippi 1968, 123), the most important one, however, is Amaterasu's grandson Ninigi who married the daughter of the Earthly Deity Ohoyamatsumi (Philippi 1968, 144–147). On the other hand, none of the Earthly Deities won a bride who belonged to the Heavenly Deities. This would be difficult to achieve, since although the Japanese myths contain some accounts of Heavenly Deities descending to Earth, it seems to have been impossible for Earthly Deities to ascend to the Plain of High Heaven (again, with the possible exception of Susanowo).

Comparative-Historical Approaches

These remarkable parallels between mythologies which are separated by a long distance in space and in time cannot be explained by the historical hypothesis introduced above with regard to the Japanese case, and which has also been proposed for the other mythologies (cf. Dumézil 1973, 11–12; Hard 2004, 24–35; Mondi 1990, 168). There are two major theories which take the comparative evidence into account: Georges Dumézil's trifunctional system and Michael Witzel's Laurasian theory.

I will start with the more recent theory proposed by Witzel. He claims that the fundamental similarities between most of the world's mythologies can only be explained by common origin. Thus, he reconstructs a basic "story line" containing mythic themes which are widely distributed over Eurasia, North Africa, Polynesia and the Americas (he calls this region by the geological term "Laurasia") but cannot be found in Australia, New Guinea and sub-Saharan Africa ("Gondwana").[3] Witzel argues that the Gondwana system of mythology, which does not contain his reconstructed "story line" "must have been that of our African ancestors: a small group of them ventured 'out of Africa' some 65,000 years ago and followed [. . .] the coastline of the Indian Ocean via Arabia, India, and Sunda Land to Australia. They became the ancestors of all non-African people. A subset of them developed the Laurasian mythological system that became increasingly dominant after the last two ice ages, some 50,000 and 20,000 years ago." (Witzel 2012, xi) Witzel's "Laurasian story line", which must be understood as a heuristic reconstruction of the mythological system that developed out of the Gondwana one, starts with the emergence of the world and contains a number of topics, which, taken together, express an ancient worldview: "the universe is ultimately regarded as a living body, not surprisingly in analogy to the human one: it is born, grows, and finally dies" (Witzel 2012, 54–55). The most important individual topics are:

> "[1] primordial waters/chaos/nonbeing/egg/giant/hill or island; [2] Father Heaven, Mother Earth, and their children (four generations/ages); [3] the pushing up of heaven; [4] incest between Heaven and his daughter; [5] revealing the light of the hidden sun; [6] *the current gods defeating their predecessors*; [7] killing the dragon; [8] the Sun deity as the father of humans (especially of chieftains); [9] the first humans and first evil deeds; [10] the origin of death/the flood; [11] heroes and nymphs; [12] the bringing of fire/food/sacred drink and so on by a culture hero; [13] the spread of humans and later, in actual, if legendary, history, of local noble (subsequently, royal) lineages and the beginning of local history; [14] the final destruction of humans, the world, and the gods; and sometimes [15] the hope for a new heaven and a new earth" (Witzel 2012, 76, my emphasis).

The episode discussed in this chapter is included as the sixth item in Witzel's list (the current gods defeating their predecessors). He explains it in the following way: "The primordial deities (Father Heaven/Mother Earth) have two sets of children: the Titans and the Olympians, to use the Greek names in the following discussion. The 'demonic' Titans (Kronos etc.) take the same genealogical and functional position in the evolution of the gods as the Germanic giants, the Japanese Kuni.no Kami (Mundane deities) [or Earthly Deities]. They oppose the 'Olympian' gods, such as Zeus, the German Æsir gods of Asgard/Valhalla, or the Japanese Ama.no Kami (Heavenly deities). [. . .] In sum, we can establish an old Laurasian myth about the succession of the several generations of deities." (Witzel 2012, 161) Witzel explicitly denies the possibility of a historical explanation: "In various mythologies we have the event, called a '(land-)ceding' process in Japan, taking place between two groups of gods, such as the Greek Titans and Olympians or the Indian Asura and Deva. This competition between cousins [. . .] is built into Laurasian mythology, and it is not one instigated by mysterious earlier settlers or Aborigines and their religion" (Witzel 2012, 162).

---

[3]Witzel's use of the terms "Laurasia" and "Gondwana" differs somewhat from common usage. For a map, see Witzel (2012), 18. His usage of the terms does not imply that his reconstructed mythologies were contemporaneous with the existence of the two supercontinents (Witzel 2012, 4–5, 443, n. 25).

A somewhat similar approach has already been suggested by Georges Dumézil (1898–1986). Like Witzel, he was inspired by historical linguistics and tried to reconstruct the mythologies corresponding to specific language families—in Dumézil's case Indo-European. "The Indo-Europeans are a set of linguistically related peoples, not a race. The term Indo-European refers in the first place to a family of languages that is spread over an area extending from Central Asia and India to Western Europe. From Europe, Indo-European languages were dispersed to many other areas of the world. All Indo-European languages of different areas and periods share a common origin." (Oosten 1985, 12) As a matter of fact, Indo-European mythology is also incorporated into Witzel's Laurasian theory as one of several intervening layers between his reconstructed Laurasian story line and attested mythologies like, say, Hesiod's *Theogony*: "The several successive layers of descendants of the original ancestor that give rise to further 'generations' are always indicated by individually developed *innovations* ("mutations"), by which they are distinguished from the earlier ancestor generation(s). These successive layerings bridge the gap between original Proto-Laurasian mythology and its oldest written (or otherwise recorded) forms" (Witzel 2012, 75).

These similarities led Nick Allen (2014, 101) to the assessment that Witzel's "common origin approach to world mythology is merely an application on a larger scale of the approach taken by Georges Dumézil (and others) to Indo-European mythology." In contrast to Witzel, however, Dumézil was not interested in reconstructing a common narrative scheme underlying the various Indo-European mythologies but rather tried to retrieve an idealized social structure which, he postulated, all Indo-European cultures shared. Dumézil believed that all early Indo-European societies were characterized "by a hierarchically ordered, tripartite social organization, each stratum of which was collectively represented in myth and epic by an appropriate set of gods and heroes." (Littleton 1973, 4) All of these three strata were linked by Dumézil with the functions they performed for society. "The first or most important function (i.e., the priestly stratum and its mythical representations) was concerned with the maintenance of magico-religious and juridical sovereignty or order; the second function (i.e., the warrior stratum and its representations) was concerned with physical prowess; and the third or least important function as far as the Indo-Europeans were concerned (i.e., the herder-cultivator stratum and its representations) was charged with the provision of sustenance, the maintenance of physical well-being, plant and animal fertility, and other related activities." (Littleton 1973, 5) It is important to note that this social structure might never have existed in reality, "but as an ideological reality has proven for the bearers of these traditions to be 'good to think with,' or even good to model social life after." (Nagy 1990, 201) According to this theory, the rivalling groups of gods in Indo-European mythologies actually represent different functions (cf. Dumézil 1973; Haugen 1967; Oosten 1984, 1985).

In this instance, however, the line was not drawn between giants and gods in Scandinavian mythology, but rather between two different groups within the gods: the Aesir and Vanir.[4] "In a far distant past the two divine groups lived at first separately, as neighbors, then they fought a fierce war, after which the most distinguished Vanir were associated with the Æsir" (Dumézil 1973, 7). According to Dumézil, the Aesir represent the first and second functions whereas the Vanir represent the third (Dumézil 1973; Nagy 1990, 217). The Greek case is somewhat more difficult to analyse within Dumézil's theoretical framework since Greek mythology was heavily influenced by Near-Eastern ideas—in Dumézil's words "Greek mythology escapes Indo-European categories" (Dumézil 1973, 37; cf. Mondi 1990, 187).[5]

Dumézil's reservations on the applicability of his model to non-Indo-European cultures notwithstanding, his theory was from the 1960s applied to Japanese mythology by Yoshida (1977, 2004, 2006) and Ōbayashi (1975, 1977), who claimed that the Indo-European ideology was brought to the Korean peninsula by the Scythians and from there advanced into the Japanese archipelago (Yoshida 2006). According to Ōbayashi (1977, 119–121), the Heavenly Deities represent "the first (sovereign) and the second (warrior) functions in Dumézilian terms", while the "social functions fulfilled by the earthly deities consist primarily in being autochthons or lords of the earth, that of being food-producers and that of insuring fertility, functions which can be subsumed under Dumézil's concept of the third function."[6]

Quantitative Approaches

Both of the hypotheses introduced in the preceding section are based on the observation that the different mythologies under consideration have a similar structure in common—be it a narrative structure (Witzel) or a social one (Dumézil). However, until now it was not possible to quantify, how similar the structures mirrored in the individual tales actually are. If one wants to prove a theory like the ones proposed by Dumézil or Witzel, it is almost impossible to avoid analysing

---

[4]Witzel (2012, 161) describes the war between the Aesir and Vanir as a variation of the theme "the current gods defeating their predecessors" which can be compared to the Indian Asuras and Devas: "two moieties in constant competition who nevertheless also cooperate periodically."

[5]It has been pointed out by Nagy (1990, 205) that "This impression of the relatively non-Indo-European nature of Greek culture and myth remains, however, an impression rather than a fact. [. . .] Their protestations of the apparently slim pickings in Greek tradition notwithstanding, Dumézil and other scholars have already built up an impressive dossier of Greek attestations of mythological patterns that can be found in other Indo-European mythologies as well."

[6]A somewhat more complex interpretation has been proposed by Hirafuji (2004, 103–175), who claims that the trifunctional order in the Japanese pantheon was only established *after* the cession of the land. Unfortunately she does not elaborate what this means for the relationship between Japanese and Indo-European mythology.

the myths in a preconceived, selective way: the importance of facts which support the thesis is exaggerated, while facts which contradict the theory tend to be passed over in silence.[7] The application of quantitative approaches could be the solution to some of these problems.

Pádraig Mac Carron and Kenna (2012, 2014) have shown that network theory can be meaningfully applied in the comparative study of mythological texts. They have compared the social networks of fictional texts with real social networks and identified markers for artificiality: "our initial research indicates that degree distributions, disassortativity and robustness are indicators which may distinguish some fictional social networks from real ones." (Mac Carron and Kenna 2014, 132) In the next step, the social networks of epic texts—the *Iliad*, *Beowulf* and the *Táin Bó Cúailnge*—were compared to the clearly fictional and to the real social networks. It was shown that "of the three narratives, only the *Iliad* turns out to be assortative. *Beowulf* is mildly disassortative and the *Táin* is very disassortative. These results indicate that the social-network properties of the *Iliad* are similar to those of real social networks and dissimilar to random networks and the fictitious social networks listed above. This is what one might expect if the narrative is a reasonable portrayal of a real society." Although this "does not categorically *prove* that the *Iliad* is based on real events", "it may be interpreted as evidence to *support* the case for historicity" (Mac Carron and Kenna 2014, 135, emphasis in the original).

In a similar vein, network theory could be fruitfully applied to the problem under discussion in this chapter: if an analysis of the *Kojiki*, the *Theogony* and the *Prose Edda* would yield social networks which are similar to real ones, this might support the historical interpretation of the struggle between the gods. If, on the other hand, the resulting networks turn out to be dissimilar to real networks but similar to each other, the comparative interpretations would gain more plausibility. It would also be interesting to observe how the interactions between the two groups are depicted in the social networks: will the various groups be recognizable as separate clusters? Are the mediating roles of Susanowo and Loki, mentioned above, discernible in the networks? If so, is there a similar figure in the *Theogony*? Of course the resulting networks could also turn out to be dissimilar to real ones and dissimilar to each other. In this case, it might be necessary to look for new answers. Such a result could also be interpreted in favour of Dumézil's reservations to consider Greek mythology (not to speak of the Japanese one) as Indo-European. In this case, the

---

[7]Nagy (1990, 202) points out that "It would be simple enough to discern the system of the three functions in any mythology, Indo-European or not, by simply picking and choosing the appropriate mythological terms at large" and criticizes that many scholars inspired by Dumézil "have engaged in precisely such dubious eclecticism". Oosten, on the other hand, criticizes Dumézil's approach itself by claiming that "Dumézil's tripartition tends to simplify the complex structure of the Indo-European pantheon", since "the Indo-European gods are complex figures" who "cannot be reduced to one particular function." (Oosten 1985, 19, 26–27) Witzel (2012, xii) is well aware that "it is impossible for any one person to have sufficient command over the languages and the intricacies of the ancient and modern texts involved" in his project and that his work "may contain some misjudgments caused by lack of familiarity."

comparison of further clearly Indo-European mythologies might be necessary to put Dumézil's hypothesis to the test. Given the exalted importance of the genealogies in the three works under discussion, it might also be interesting to draw networks of the marital relations between the deities apart from the "friendly" and "hostile" networks created by Mac Carron and Kenna (2014, 134) in their study.

Since both of the comparative hypotheses introduced above explain the parallels between different mythologies with common origin, phylogenetic analysis is another method which might be employed to counter-check their results. "Phylogenetics was originally developed to investigate the evolutionary relationships among biological species, and has become increasingly popular in studies of cultural phenomena"; "the aim of a phylogenetic analysis is to construct a tree or graph that represents relationships of common ancestry inferred from shared inherited traits (homologies)." (Tehrani 2013, 2) As a matter of fact, Witzel heavily relies on recent findings in genetic studies in reconstructing the route taken by the populations who distributed the Laurasian story line over large portions of the world (Witzel 2012, 207–241). Moreover, his approach is inspired by phylogenetics as becomes apparent in the following quote: "Historical comparative studies have established pedigrees of the development of *Homo sapiens*, of most human languages, and of human DNA [. . .] from the 'African Eve' to that of all modern humans. In all cases, the descendants of a certain parent along the line of descent are characterized by *common mutations* or as called in linguistics, by *shared innovations*. The same comparative and historical method will be used in this book for the reconstruction of earlier forms of mythology" (Witzel 2012, 47).

Witzel does not, however, employ quantitative methods of phylogenetics.[8] Such methods have recently been successfully applied to folklore studies (Tehrani 2013) and to mythology (d'Huy 2013). Until now, these studies were limited to single tale types (Little Red Riding Hood, Pygmalion). Considering the fact that Jamshid Tehrani analysed 72 plot variables in 58 variants of tales belonging to the types of Little Red Riding Hood (ATU 333) and The Wolf and the Kids (ATU 123) (Tehrani 2013, 3), an analysis of comparable detail seems unrealistic for the whole story line reconstructed by Witzel. It would be possible to make similar analyses of particular tales which are grouped together by Witzel and used as basis for the reconstruction of particular elements (or themes) in the story line; however, this would amount to a departure from Witzel's method of comparing whole mythologies instead of particular motifs or mythemes (Witzel 2012, 15). Alternatively, individual works—like the *Theogony* or *Edda*—might be searched for the individual themes of Witzel's reconstructed story line which might then serve as the plot variables (presence/absence, sequence of themes). Such an approach,

---

[8]This might be due to the fact that Witzel models his approach mainly on historical linguistics, a field which differs, for example, from Biology in that it relies more on qualitative assessments which are "largely based on an individual researcher's knowledge of multiple languages and a system of constructing sound correspondences among those languages", whereas biologists "have typically applied quantitative analytical approaches that implement specific algorithms" (Nunn 2011, 39).

however, would involve a high degree of subjectivity: how similar must particular episodes be in order to be classified as the same theme?

A third quantitative approach that has been applied to the comparative study of myth and folklore by Yuri Berezkin is principal component analysis. This statistical method brings out dissimilarity patterns in databases.[9] Berezkin has built up an enormous dataset of international folklore motifs (http://www.ruthenia.ru/folklore/berezkin). Based on these data, he creates maps of the geographical distribution of individual motifs (e.g. Berezkin 2014, 284, fig. 1). In a further step, he compares the inventory (presence/absence) of motifs in different traditions. In this way, clusters of similar traditions and their geographic distribution emerge (Berezkin 2015). From these maps, we can infer the (approximate) centres of diffusion of individual motifs but also of particular inventories of motifs. Obstacles of this method for the purpose of the present study are (1) that the sequence of motifs is not taken into consideration, which makes it difficult to use for structural approaches, and (2) that the resultant maps lack a diachronic dimension, which makes it impossible to trace the historical development of certain motifs or combinations of motifs. However, it can be used to countercheck Dumézil's and Witzel's hypotheses by showing whether or not Indo-European or Laurasian clusters emerge on the geographic distribution maps.

In conclusion, it should be emphasized that the quantitative methods introduced in the last section will not replace traditional approaches to mythology. They might, however, bring a certain degree of measurability and objectivity to the often highly speculative field of comparative mythology. (Although it should be kept in mind that quantitative approaches, too, involve subjective judgements on the researcher's part, for instance, when she defines the plot variables to be analysed, decides for a specific tool to create a phylogeny, or classifies the links in a social network, etc.) It seems a reasonable first step, to start from established theories like the ones proposed by Dumézil or Witzel—or the Finnish historic-geographic school as done by Tehrani (2013). As soon as a bigger pool of quantitative analyses is available and the potential of the approach is recognized in the humanities, more ambitious problems can be tackled.

References

Allen, N. J. (2014). Comparing mythologies on a global scale: Review article of E. J. Michael Witzel, The origins of the world's mythologies. *Journal of the Anthropological Society of Oxford, 6*(1), 99–103. https://www.isca.ox.ac.uk/fileadmin/ISCA/JASO/2014/Allen.pdf.
Antoni, K. (Ed.). (2012). *Kojiki: Aufzeichnung alter Begebenheiten*. Berlin: Verlag der Weltreligionen.

---

[9]For an accessible introduction, see Cavalli-Sforza and Cavalli-Sforza (1995), 147–150 and Stone et al. (2007), 122–125; cf. Witzel (2012), 37–41.

Aston, W. G. (Trans.) (1956 [1896]). *Nihongi: Chronicles of Japan from the earliest times to A.D. 697*. London: George Allen & Unwin Ltd.

Bascom, W. (1984). The forms of folklore: Prose narratives. In A. Dundes (Ed.), *Sacred narrative: Readings in the theory of myth* (pp. 5–29). Berkeley, Los Angeles, London: University of California Press.

Berezkin, Y. (2014). Native American and Eurasian elements in post-Columbian Peruvian tales. In K. Antoni & D. Weiß (Eds.), *Sources of mythology: Ancient and contemporary myths. Proceedings of the Seventh Annual International Conference on Comparative Mythology (15–17 May 2013, Tübingen)* (pp. 281–302). Religionswissenschaft: Forschung und Wissenschaft 12. Zürich: Lit.

Berezkin, Y. (2015). Spread of folklore motifs as a proxy for information exchange: Contact zones and borderlines in Eurasia. *Trames, 19*(1), 3. doi:10.3176/tr.2015.1.01.

Byock, J. L. (Ed.). (2005). *The Prose Edda: Norse mythology*. London: Penguin.

Caldwell, R. S. (Ed.). (1987). *Hesiod's Theogony: Translated, with introduction, commentary, and interpretive essay*. Cambridge, MA: Focus classical library, Focus Information Group.

Cavalli-Sforza, L. L., & Cavalli-Sforza, F. (1995). *The great human diasporas: The history of diversity and evolution*. Reading, MA: Helix books Addison-Wesley.

Chamberlain, B. H. (Trans.) (1982 [1882]). *The Kojiki: Records of ancient matters*. Rutland, Tōkyō: Charles E. Tuttle Company.

Csapo, E. (2005). *Theories of mythology*. Malden: Oxford, Victoria: Blackwell.

d'Huy, J. (2013). A phylogenetic approach to mythology and its archaeological consequences. *Rock Art Research, 30*(1), 115–18.

Doty, W. G. (1986). *Mythography: The study of myths and rituals*. Tuscaloosa, AL: University of Alabama Press.

Dumézil, G. (Ed.) (1973). *Gods of the ancient northmen*. UCLA Center for the Study of Comparative Folklore and Mythology, Publications 3. Berkeley: University of California Press.

Florenz, K. (Trans.) (1901). *Japanische Mythologie: Nihongi "Zeitalter der Götter" nebst Ergänzungen aus andern alten Quellenwerken*. Mitteilungen der Deutschen Gesellschaft für Natur- und Völkerkunde Ostasiens Supplementband 4. Tōkyō: Hobunsha.

Hard, R. (2004). *The Routledge handbook of Greek mythology: Based on H. J. Rose's Handbook of Greek mythology*. London, New York: Routledge.

Haugen, E. (1967). The mythical structure of the ancient Scandinavians: Some thoughts on reading Dumézil. In *To Honor Roman Jakobson: Essays on the occasion of his seventieth birthday* (pp. 855–868). The Hague, Paris: Mouton.

Hyde, L. (2010). *Trickster makes this world: Mischief, myth, and art*. New York: Farrar, Straus and Giroux.

Hirafuji, K. (2004). *Shinwagaku to Nihon no kamigami*. Tōkyō: Kōbundō.

Littleton, C. S. (1973). *The new comparative mythology: An anthropological assessment of the theories of Georges Dumézil*. Berkeley, Los Angeles, London: University of California Press.

Mac Carron, P., & Kenna, R. (2012). Universal properties of mythological networks. *Europhysics Letters, 99*(2), 28002. doi:10.1209/0295-5075/99/28002.

Mac Carron, P., & Kenna, R. (2014). Network analysis of *Beowulf*, the *Iliad* and the *Táin Bó Cúailnge*. In K. Antoni & D. Weiß (Eds.), *Sources of mythology: Ancient and contemporary myths. Proceedings of the Seventh Annual International Conference on Comparative Mythology (15–17 May 2013, Tübingen)* (pp. 125–141). Religionswissenschaft: Forschung und Wissenschaft 12. Zürich: Lit.

Matsumura, T. (1951). Dōhōjin no minzoku-bunkashiteki kōsatsu. *Minzokugaku kenkyū, 16*(2), 1–28.

Matsumura, T. (1954–1958). *Nihon shinwa no kenkyū* (4 vols). Tōkyō: Baifūkan.

Mondi, R. (1990). Greek mythic thought in the light of the Near East. In L. Edmunds (Ed.), *Approaches to Greek myth* (pp. 142–98). Baltimore: Johns Hopkins University Press.

Nagy, J. F. (1990). Hierarchy, heroes, and heads: Indo-European structures in Greek myth. In L. Edmunds (Ed.), *Approaches to Greek myth* (pp. 200–238). Baltimore: Johns Hopkins University Press.

Nunn, C. L. (2011). *The comparative approach in evolutionary anthropology and biology.* Chicago: University of Chicago Press.

Ōbayashi, T. (1975). *Nihon shinwa no kōzō.* Tōkyō: Kōbundō.

Ōbayashi, T. (1977). The structure of the pantheon and the concept of sin in ancient Japan. *Diogenes, 25,* 117–132.

Oka, M. (2012). *Kulturschichten in Alt-Japan,* ed. Josef Kreiner (2 vols). JapanArchiv: Schriftenreihe der Forschungsstelle Modernes Japan 10. Bonn: Bier'sche Verlagsanstalt.

Oosten, J. G. (1984). The war of the gods in Scandinavian mythology. In H. G. Kippenberg (Ed.), *Struggles of gods: Papers of the Groningen work group for the study of the history of the religions* (pp. 193–223). Religion and Reason 31. Berlin, New York: Mouton.

Oosten, J. G. (1985). *The war of the gods: The social code in Indo-European mythology.* International Library of Anthropology. London, Boston, Melbourne, Henley: Routledge & Kegan Paul.

Philippi, D. L. (Trans.) (1968). *Kojiki.* Tōkyō: University of Tōkyō Press.

Piggott, J. R. (1989). Sacral kingship and confederacy in early Izumo. *Monumenta Nipponica, 44*(1), 45–74.

Stone, L., Lurquin, P. F., & Cavalli-Sforza, L. L. (2007). *Genes, culture, and human evolution: A synthesis.* Malden, MA: Blackwell.

Takagi, T. (1973). Susanowo no mikoto shinwa ni arawaretaru takamgahara yōso to Izumo yōso. In T. Ōbayashi (Ed.), *Zōtei Nihon shinwa densetsu no kenkyū* 1, 208–56. Tōyō bunko 241. Tōkyō: Heibonsha.

Tehrani, J. J. (2013). The phylogeny of little red riding hood. *PLoS One, 8*(11), e78871. doi:10.1371/journal.pone.0078871.

Tsuda, S. (1963). *Nihon koten no kenkyū: Jō.* Tsuda Sōkichi zenshū 1. Tōkyō: Iwanami Shoten.

van Baaren, T. (1984). A few essential remarks concerning positive and negative relations between gods. In H. G. Kippenberg (Ed.), *Struggles of gods: Papers of the Groningen work group for the study of the history of the religions* (pp. 7–11). Religion and Reason 31. Berlin, New York: Mouton.

de Vries, J. (1956–1957). *Altgermanische Religionsgeschichte.* 2nd ed. 2 vols. Grundriss der germanischen Philologie 12. Berlin: Walter de Gruyter.

Wittgenstein, L. (1953). *Philosophical investigations.* G. E. M. Anscombe (Trans.). Oxford: Basil Blackwell.

Witzel, M. (2012). *The origins of the world's mythologies.* Oxford, New York: Oxford University Press.

Yoshida, A. (1977). Japanese mythology and the Indo-European trifunctional system. *Diogenes, 25,* 93–116.

Yoshida, A. (2004). Nihon shinkai no sankinōteki kōzō. In T. Ōbayashi (Ed.), *Susanoo shinkō jiten* (pp. 217–230). Shinbutsu shinkō jiten shirīzu 7. Tōkyō: Ebisu kōshō.

Yoshida, A. (2006). Dumezil and comparative studies of Eurasian myths. In T. Osada & N. Hase (Eds.), *Proceedings of the Pre-Symposium of RHIN and 7th ESCA Harvard-Kyoto Roundtable* (pp. 236–242). Kyōto: Research Institute for Humanity and Nature.

Printed in the United States
By Bookmasters